Managing environmental pollution

Managing Environmental Pollution presents a comprehensive introduction to the nature of pollution, its impact on the environment, and the practical options and regulatory frameworks for pollution control. Sources of pollution, regulatory controls including the role of authorities and precautionary and polluter pays principles, technological solutions, management and mitigation techniques and assessment tools, are examined in each key area: air, freshwater and marine pollution, contaminated land and radioactive substances. Illustrated with a wide range of case examples from the UK, Europe, North America and world-wide, this book offers an invaluable up-to-date guide to both the principles and practice of pollution management.

Andrew Farmer is a Fellow of the Institute for European Environmental Policy. Educated at Oxford and York Universities, he undertook aquatic ecological research at the Universities of St Andrews, Florida and Wisconsin, before joining the air pollution research group at Imperial College. Following this he spent six years as a pollution specialist with English Nature.

Routledge Environmental Management Series

This important series presents a comprehensive introduction to the principles and practices of environmental management across a wide range of fields. Introducing the theories and practices fundamental to modern environmental management, the series features a number of focused volumes to examine applications in specific environments and topics, all offering a wealth of real-life examples and practical guidance.

MANAGING ENVIRONMENTAL POLLUTION
Andrew Farmer

COASTAL AND ESTUARINE MANAGEMENT
Peter W. French

Forthcoming titles:

ENVIRONMENTAL IMPACT ASSESSMENT
A. Nixon

WETLAND MANAGEMENT
L. Heathwaite

COUNTRYSIDE MANAGEMENT
R. Clarke

Managing environmental pollution

Andrew Farmer

London and New York

First published 1997
by Routledge
11 New Fetter Lane, London EC4P 4EE

Simultaneously published in the USA and Canada
by Routledge
29 West 35th Street, New York, NY 10001

Typeset in Ehrhardt by Keystroke, Jacaranda Lodge, Wolverhampton
Printed and bound in Great Britain by Redwood Books, Trowbridge, Wiltshire

British Library Cataloguing in Publication Data
A catalogue record for this book is available from the British Library

Library of Congress Cataloguing in Publication Data
A catalogue record for this book has been requested

ISBN 0–415–14514–7 (hbk)
ISBN 0–415–14515–5 (pbk)

Contents

List of illustrations   viii
Preface   xiii
Acknowledgements   xv
Abbreviations and acronyms   xvi

1 Introduction and basic principles   1

Introduction   2
What is pollution?   2
Sustainable development   4
Carrying capacity   5
The management of environmental risks: using the precautionary
   principle   6
Setting standards   9
Polluter pays and user pays   11
Economic analysis   13
Monitoring   16
Chapter summary   18
Recommended reading   19

2 Air pollution and its impacts   21

Introduction   22
Air pollutants and their sources   22
Air pollution impacts on health   34
Impacts of air pollution on the natural environment   40

Impacts of air pollution on the built environment   51
Impacts on visibility   53
Economic impacts   54
Chapter summary   55
Recommended reading   56

3 Air pollution, regulation and management   57

Introduction   58
Air pollution regulation in the United Kingdom   58
Air quality standards in the United Kingdom   63
Air pollution regulation by the European Union   70
The United Nations Economic Commission for Europe (UNECE)   75
The critical loads approach: assessing national and international policies   77
Monitoring air pollution   79
Modelling air pollution dispersion   82
Techniques to reduce air pollution   87
Environmental assessment   99
Chapter summary   100
Recommended reading   101

4 The freshwater environment   103

Introduction   104
Threats to freshwater   104
United Kingdom and European Union legislative background   105
A strategy for water quality in England and Wales   109
Sources of pollution in freshwater   112
Groundwater   149
Chapter summary   151
Recommended reading   152

5 Pollution of the marine environment   155

Introduction   156
Types of marine pollution and their impacts   156
Managing marine pollution   182
Chapter summary   186
Recommended reading   187

6 Managing pollution from radioactive substances   189

Introduction   190
What are radioactive substances?   190
Measuring radiation   191

Sources of man-made radiation   192
The health effects of radiation   193
The effect of radionuclides in the wider environment   195
Radionuclides in the marine environment   203
Regulation   205
The peculiar case of radon gas   208
Chapter summary   211
Recommended reading   212

7 People, politics and the press   213

Introduction   214
The public   215
The media   219
Politicians   221
Industry   222
Non-governmental bodies   223
The anti-environmental backlash   224
Chapter summary   225
Recommended reading   225

Glossary   227
References   230
Index   240

Illustrations

Plates

2.1 Smogs occurring in hot, dry conditions: Sydney   27

2.2 Growing plants in Solardomes to examine the effect of air pollution   43

3.1 A typical British coal-fired power station   59

4.1 Lower Swansea valley: heavy-metal contamination of soils and waters from industrial activity   117

4.2 Examining the effects of acid rain on catchments by roofing the entire catchment: Gardsjon, south-west Sweden   140

4.3 Liming to reduce acidification of freshwaters: Sweden   141

4.4 Liming forest soils: Sweden   142

4.5 Liming using a helicopter: Sweden   143

5.1 Tropical mangrove habitats   168

6.1 Arctic lichen community   202

7.1 Spanish protest against nuclear power   217

Figures

2.1 The distribution of acid soils in Great Britain   41

2.2 A diagrammatic representation of the possible mechanisms of forest damage that may be caused by acid deposition   42

2.3 A mechanistic model of the soil and plant processes affected by nitrogen deposition   45

2.4 The distribution of 2,164 Central European plant species across a gradient of nitrogen indicator values, showing that threatened species predominate in nitrogen-poor situations   46

2.5   Air pollution zones in Britain, as defined by the distribution of lichens   50
3.1   The distribution of the UK automatic air quality monitoring stations
      in 1994   80
3.2   The emission rates for different pollutants for non-catalytic cars driving
      at different speeds   97
4.1   Changes in total phosphorus, soluble phosphorus, chlorophyll a and
      Daphnia in Cockshoot Broad, Norfolk   135
4.2   Changes in lake pH for a range of Swedish lakes over thirty to forty
      years   137
4.3   Changes in pH at Lake Gardsjon, Sweden, over the last 100 years, as
      determined by changes in diatom populations   140
4.4   The predicted effect of distance from a target area on mortality among
      four aquatic invertebrates, following aerial spraying with the pesticide
      Deltamethrin: *Asellus aquaticus, Gammarus pulex, Sigara dorsalis* and
      Centroptilum pennulatum   148
5.1   A diagram of the environment flows of nitrogen within the European
      Union in 1988   181
6.1   The dispersion of the plume and reported arrival times of detectable
      activity in the air, following the Chernobyl accident in 1986   199

Tables

1.1   Costs of discharges (in Ecu in 1992) for five EU member states for
      two types of discharge, and the difference in costs between the two   13
1.2   Examples of economic values assigned by different workers to health
      conditions   15
1.3   Estimated costs to human health of road transport in 1993 in the UK   15
1.4   The external cost of road transport in the UK, by fuel type   16
2.1   Estimates of the percentage contribution to total UK annual ammonia
      emissions from two studies   24
2.2   Percentage of aerial emissions of pollutants from motor vehicles, by
      vehicle type, in the UK in 1990   25
2.3   UK 1990 emissions of pollutants from motor vehicles and the range of
      projected emissions for the year 2025, high and low forecasts   25
2.4   Estimated emissions of nitrogen oxides for different emission sources
      in the UK   27
2.5   The percentage contribution to PM_{10} emissions in the UK by different
      sources   30
2.6   UK emissions of sulphur dioxide, by source   31
2.7   Estimated emissions of volatile organic compounds in the UK in 1990 and
      1980   33
2.8   Estimates of the increase in occurrence of respiratory problems in
      asthmatic patients with changing concentrations of PM_{10} 37

2.9   Lowest observable adverse effect levels for health effects of lead in adults   39

2.10   Lowest observable adverse effect levels for health effects of lead in children   39

2.11   Changes in defoliation of selected tree species between 1988 and 1991, resulting from surveys across twenty-eight European countries under the International Cooperative Programme on Assessment and Monitoring of Air Pollution Effects on Forests   43

2.12   Examples of the types of impacts that have been recorded on vegetation and communities for different dusts   49

2.13   Damage caused to different types of material by different air pollutants   52

2.14   The present value of economic benefits resulting from the UNECE sulphur dioxide control strategy for the UK   54

3.1   Standards recommended for the UK by the Expert Panel on Air Quality Standards together with the proposed government targets to achieve these standards   64

3.2   Examples of international air quality standards for human health outside the UK   65

3.3   Critical levels of air pollutants   66

3.4   Area and number of statutory nature conservation sites in Britain that remain at risk from acidification under four sulphur emission reduction projections, from a 1980 baseline   79

3.5   Number of sites for different pollutant types in the UK National Automatic Monitoring Network by the end of 1996   82

3.6   Advantages and disadvantages of different air quality monitoring techniques   83

4.1   Maximum Admissible Concentrations for drinking water for a range of selected contaminants as defined in the EC Drinking Water Directive   107

4.2   Main issues addressed in a Catchment Management Plan   111

4.3   Summary of the sources of pollution incidents recorded in England and Wales during 1993   113

4.4   Examples of the range of diseases transmitted in infected waters, according to type of organism and symptoms caused   114

4.5   Concentrations of metals found in the tissues of aquatic bryophytes   118

4.6   The biological oxygen demand for different sources of organic pollution   119

4.7   Percentage removal of pollutants by two treatment methods compared with untreated motorway run-off   121

4.8   Changes in nitrate concentrations in four English rivers from 1928 to 1976   122

4.9   OECD categories for trophic status based on the total phosphorus concentration in the water column   124

4.10   Characteristics of freshwaters of different trophic status based on OECD categories   125

4.11 The number of SSSIs in England showing symptoms of
      eutrophication   131
4.12 The primary cause of eutrophication in seventy-nine SSSIs in
      England   132
4.13 Primary management requirements for the seventy-nine SSSIs affected
      by eutrophication   133
4.14 The Special Ecosystem target levels for phosphate in freshwaters   133
4.15 Fish and invertebrate status of Welsh streams, including the mortality
      of caged trout   138
4.16 Effects of decreasing the degree to which critical loads for freshwaters
      in Norway are exceeded, on the quantity of lime used to ameliorate
      acidification and its cost (1995 equivalent)   142
4.17 Types of remediation activity undertaken to control pollution to
      groundwaters from 1,205 point sources in England and Wales   150
5.1   The relative quantities of different major sources of global marine
      pollution   157
5.2   Processes in the UK producing heavy-metal pollutants   159
5.3   Loading of heavy metals (tonnes/year) to the sea from discharges
      from England and Wales in 1985 and 1993 and the percentage
      reduction achieved   160
5.4   Environmental Quality Standards (EQSs) for heavy metals in the
      marine environment   162
5.5   Major loadings of metals to the Humber estuary from sewage and
      industrial discharges in 1990   164
5.6   Ranges in the average concentrations of heavy metals in the Humber
      tidal rivers, the Humber estuary and in estuarine sediments between
      1980 and 1990   164
5.7   A list of the major organic toxins considered to be of most importance
      in UK marine waters   165
5.8   The relative contribution of oil from different sources   167
5.9   Estimated total inputs to the marine area for 1995 covered by the Oslo
      and Paris Conventions (north-east Atlantic) from municipal treatment
      plants and agriculture of signatories to the conventions, and the effects
      of planned and executed pollution control measures   177
5.10 Estimated total inputs to the marine area for 1995 covered by the Oslo
      and Paris Conventions (north-east Atlantic) from industry and
      atmospheres of signatories to the conventions, and the effects of
      planned and executed pollution control measures   178
5.11 Summary of the relative importance of different sources of inorganic
      nutrients to the Scottish North Sea Coastal Zone   179
6.1   The percentage exposure to current global radiation from different
      sources, as compared with natural sources   192
6.2   Ranges of acute lethal doses for different taxonomic groups   195
6.3   Vegetation zonation from the centre of the test area for above-ground
      tests at the Semipalatinsk nuclear test site   197

6.4 The relative impact of Chernobyl radiation on forests immediately
around the accident site   201
6.5 Types of sampling undertaken in routine monitoring in the marine
environment for radionuclides in the UK   204
6.6 Calculated critical group doses for the major nuclear sites in England
assuming releases are at 100 per cent of those which have been
authorised   208
6.7 Differences in the percentage contribution of different radiation
sources to the total dose of radiation for different types of people
living in the UK   210

Boxes

2.1 Units of measurement   23
4.1 The black and grey lists of substances from the European Union
Dangerous Substances Directive   106
4.2 Biological monitoring   115
5.1 International frameworks for protecting marine areas from pollution
(and other damaging activities)   183
5.2 Direct Toxicity Assessment   185

Preface

The management of environmental pollution is an increasingly complex subject, both scientifically and politically. A pollution manager has to be informed by a wide range of rapidly developing scientific information and yet, at the same time, has to make decisions that may be driven by social or political factors. The task is a difficult one. This volume aims to begin to introduce the reader to how environmental pollution is managed and the context in which such work is undertaken. It is aimed primarily at those who may be embarking on a career in environmental management, i.e. an undergraduate or post-graduate audience. This book views each subject from the point of view of the pollution manager, examining the scientific, regulatory and sociopolitical context in which he or she would work.

In a volume of this size it is not possible to provide a comprehensive examination of all pollution issues and their management. Some subjects are not covered at all, for example the management of pesticide use, although some reference is made to pesticide contamination in freshwaters and marine systems. Within other topics some issues have also not been able to be addressed, e.g. that of indoor air pollution, although indoor radon pollution is discussed within the context of radionuclide control as it has broader lessons for the pollution manager. It is also important to view this volume within the context of the whole series on environmental management to be published by Routledge. There is to be, for example, a separate volume covering the issue of waste, so the pollution aspects of waste disposal will, largely, not be covered here.

I have taken a 'classic' media-based approach to examining pollution issues, i.e. by undertaking a separate consideration of air, freshwater, marine and radioactive substances. This approach is, in effect, bucking the trend of integrating pollution management. However, in practice much of the information relating to each of these media is far from integrated. Some topics are, however, obviously common to more than one medium. Integrated Pollution Control, for example, covers all media, but is here introduced in Chapter 3 on air pollution. Direct Toxicity Assessment is equally applicable to freshwater and marine systems, but is covered in the marine section (Chapter 5). Many of the pollution controls that reduce discharges to rivers are as much aimed at protecting coastal waters as the rivers themselves, but are highlighted in the chapter on freshwaters. It is important, therefore, not to view one chapter in isolation from the rest of the volume.

In the space available it is not possible to go into great depth for all of the issues that are discussed. Indeed, many of them deserve whole volumes of their own. However, it is in the detail that many of the more interesting aspects of pollution management become most apparent. I have, therefore, included a wide range of case studies which illustrate many of the general points that are raised. I also considered that it would be useful to examine one area of pollution management in somewhat greater detail to illustrate more clearly the complexities of the scientific information, regulatory decision-making and other aspects of pollution management. I have, therefore, devoted two chapters to air pollution management.

Many pollution texts treat environmental management as a scientific exercise. This is, however, only partly the case – it is a necessary condition, but not sufficient in itself. I have also tried, therefore, to emphasise that the pollution manager works in the 'real world', with all the social, economic and political pressures that result. Most pollution managers are ultimately responsible to a very broad constituency. In some cases, communicating correctly the results of one's decision-making is probably nearly as important as the procedures that lead to that decision. The final chapter, therefore, argues strongly against an 'ivory tower' approach to pollution management, indicating the opportunities and threats that are posed by dealing with the public, media, and so on.

Finally, I must stress that the opinions contained within this volume are entirely my own and should not be considered as attributable to the organisation for which I work.

Andrew Farmer

Acknowledgements

Without singling out individuals I would like to thank the many colleagues I have worked with over many years, who have encouraged and inspired my interest and concern about the effects of environmental pollution.

The following are thanked for permission to reproduce the figures: the Joint Nature Conservation Committee (Figures 2.1, 2.3 and 2.4), the Controller of Her Majesty's Stationery Office (Figures 2.2 and 3.2), David & Charles (Figure 2.5), ENSIS Publishing (Figure 3.1), English Nature (Figure 4.4), Cambridge University Press (Figure 5.1), UNESCO (Figure 6.1).

Abbreviations and acronyms

ALARA   as low as reasonably achievable
ANC   acid neutralising capacity
AQMA   Air Quality Management Area
BATNEEC   best available techniques not entailing excessive cost
BOD   biological oxygen demand
BPEO   best practicable environmental option
BPM   best practicable means
CAP   Common Agricultural Policy
CFCs   chlorofluorocarbons
CRI   Chemical Release Inventory
DMS   dimethyl sulphide
DNA   deoxyribonucleic acid
DoE   Department of the Environment
DTA   direct toxicity assessment
EA   Environment Agency
EAA   European Environment Agency
EALs   environmental assessment levels
EPAQS   Expert Panel on Air Quality Standards
EQS   environmental quality standard
EU   European Union
FGD   flue gas desulphurisation
HCFCs   hydrochlorofluorocarbons
HMIP   Her Majesty's Inspectorate of Pollution
IAEA   International Atomic Energy Authority
ICRP   International Commission on Radiological Protection
IGCC   integrated gasification combined cycle

IPC	Integrated Pollution Control
IPPC	Integrated Pollution Prevention and Control
IQ	Intelligence Quotient
LAAPC	Local Authority Air Pollution Control
LD_{50}	lethal dose for 50 per cent of a population
LEAP	Local Environment Agency Plan
MAFF	Ministry of Agriculture, Fisheries and Food
MOT	Ministry of Transport
NGO	non-governmental organisation
NOEL	no effect level
NOx	nitrogen oxides
NPS	National Park Service
NRA	National Rivers Authority
NRPB	National Radiological Protection Board
NSA	nitrate sensitive area
NSCA	National Society for Clean Air
NVZ	nitrate vulnerable zone
OECD	Organisation for Economic Cooperation and Development
OPEC	Organisation of Petroleum Exporting Countries
PAHs	polyaromatic hydrocarbons
PCBs	polychlorinated biphenyls
PEC	predicted environmental concentration
PM_{10}	particles of less than 10 μm in diameter
POCP	photochemical oxidant creation potential
PSP	paralytic shellfish poisoning
RSPCA	Royal Society for the Prevention of Cruelty to Animals
SAS	Surfers Against Sewage
SCR	selective catalytic reduction
SEPA	Scottish Environmental Protection Agency
SNCR	selective non-catalytic reduction
SOx	sulphur oxides
SSSIs	Sites of Special Scientific Interest
SWQOs	statutory water quality objectives
TBT	tributyl tin
TEQ	toxic equivalent
UN	United Nations
UNECE	United Nations Economic Commission for Europe
VOCs	volatile organic compounds
WHO	World Health Organisation

Chapter 1

Introduction and
basic principles

◆ Introduction   2
◆ What is pollution?   2
◆ Sustainable development   4
◆ Carrying capacity   5
◆ The management of environmental risks:
  using the precautionary principle   6
◆ Setting standards   9
◆ Polluter pays and user pays   11
◆ Economic analysis   13
◆ Monitoring   16
◆ Chapter summary   18
◆ Recommended reading   19

Introduction

This book aims to provide an outline of some of the main problems facing a manager of environmental pollution and the types of approach that may be taken to seek solutions to these problems. The pollution manager is a key element in protecting the environment. A pollution manager is a professional who has the skills to ensure that the correct analysis is undertaken and correct practice adopted to ensure that pollutant releases do not harm the environment. Much of the work of a pollution manager, from whatever standpoint, can be fairly mundane. He or she may be performing routine chemical analysis, drafting environmental statements, modelling pollutants, etc. As a result it is easy for them to lose sight of the aim of their work. If this happens, they should take a walk, literally or metaphorically, through their local community. There may be children playing in the park, who may be at risk from vehicle emissions. The local river with its rich plant life and birds may be at risk from eutrophication. The stonework of the historic church may be blackened and crumbling from acid rain. Their protection is the task of the pollution manager – to maintain or enhance what is valued in our environment. So, whenever the work becomes mundane, it is important to remember what its goals are.

This chapter examines the nature of 'pollution' and some of the variations in meaning of that term. It also sets the work of a pollution manager in the context of sustainable development. This involves an understanding of a range of concepts, including carrying capacity, the precautionary principle and the polluter pays principle. Each of these is discussed. There is a general treatment of how pollutants enter the environment and how different organisms may be affected. This includes a need for the use of environmental standards. These are set in different ways for different purposes and this chapter will examine these. There are many issues in pollution management that cut across the different media. Rather than deal with most of them in this introductory chapter, it is better to consider them alongside the treatment of specific pollution problems so that they may be better understood. However, this chapter will make some general remarks about the role of monitoring in environmental management.

What is pollution?

There are a number of definitions of 'pollution'. The UK sustainable development strategy (DoE, 1995) defines it as:

> a substance which is present at concentrations which cause harm or exceed an environmental standard.

This definition is not sufficient, however. Thermal pollution of freshwater and marine systems, for example, is not caused by 'substances', but by energy. There is also a danger of circularity in defining it on the basis of breaching environmental standards. Standards should be set due to pollution problems, rather than pollution problems defined in relation to standards.

Some would advance a definition whereby pollution refers to any change in the environment due to human activity. However, in practical terms this fails to distinguish between innocuous behaviour and dangerous activities. It provides no guide to action for an environmental manager. It should also be noted that not all pollution is caused by human activity. Contaminated water may arise from natural ore deposits or dust may fall from volcanic activity. Perhaps the most pressing issue for some pollution managers is naturally occurring radon gas (see Chapter 6).

The above definition does include the most common component of many, i.e. a reference to 'harm'. The link between pollutant and damage is given in the 1996 EU Directive on Integrated Pollution Prevention and Control provides the following definition:

> 'pollution' shall mean the direct or indirect introduction as a result of human activity, of substances, vibration, heat or noise into the air, water or land which may be harmful to human health or the quality of the environment, result in damage to material property, or impair or interfere with amenities and other legitimate uses of the environment.

Pollutants are something in the environment that cause harm, damage, etc. Some substances occur naturally (e.g. nitrate in water or ozone in air) and it is an excessive quantity of these substances which is harmful. Others have no threshold for effect (e.g. radioactive substances or *carcinogens* such as dioxins).

There is a problem, however, in defining 'harm' or 'damage'. This incorporates issues of significance and social values. It is possible to take a strict definition and define 'harm' as almost any response (assumed to be negative) in environmental receptors (humans, vegetation, etc.). However, many pollutants at very low concentrations, will cause reactions in such receptors, e.g. a measurable physiological response. This may be entirely trivial and, therefore, 'harmless'. An alternative is to look for 'significant' effects. For those pollutants with thresholds for impact, this is amenable to scientific analysis to determine the nature of the thresholds. However, for those without thresholds, the significance is set at a level of 'acceptable' impact. There is no safe level of radionuclide exposure, for example, but an acceptable level is identified. This, of course, involves an incorporation of social values into levels of acceptable impact.

For substances released without any known impact, their presence in the environment is often termed 'contamination', to distinguish it from pollution. Thus a released substance contaminates the environment, and if it causes harm it is a pollutant. However, most members of the public would consider a 'contaminant' as harmful in common usage of the word.

While there is some confusion about the meaning of 'pollution' in the English language, this can become even more evident when examining the different meanings in different cultures. Boehmer-Christiansen and Skea (1991), for example, provide an interesting comparison of the perception of 'pollution' in Britain and Germany. In Britain 'pollution' tends to be perceived in terms of its effect on the environment, i.e. a substance that is released becomes a 'pollutant' if it has an impact in some way. However, the German word 'Verschmutzung' is much stronger than 'pollution'. It is derived from 'Schmutz', meaning 'dirt'. The distinction between the presence of a substance and its environmental impact is, therefore, blurred in this instance – the presence of 'dirt', implies an impact. This is important in understanding how each society may view the aims of pollution regulation, e.g. contrasting the reduction of emissions to 'harmless' or eliminating them altogether.

Sustainable development

The management of environmental pollution is a key element in achieving sustainable development. Sustainable development is an important concept that is often mis-understood and misinterpreted. However, it is now the overall framework in which environmental protection activities, including pollution management, are undertaken in many countries. The most widely quoted definition of sustainable development is that of the Brundtland Report (WCED, 1987):

> development that meets the needs of the present without compromising the ability of future generations to meet their own needs.

The Brundtland Report looks to a change in economic activity that is compatible with environmental protection, i.e. 'the possibility of a new era of economic growth, based on policies that sustain and expand the environmental resource base'.

The word 'sustainable' is used in many contexts and it is important to note that these do not necessarily refer to sustainable development. For example, as Europe currently emerges from recession there is considerable discussion among politicians for 'sustainable economic growth'. This means growth that can be kept going, i.e. not faltering growth leading back to recession. The general history, for example, of the twentieth century is of overall sustainable economic growth, but much of this has been the antithesis of sustainable development. Sustainable development does require a sustained economy, but this has to be achieved while sustaining the environment. Pollution managers are key individuals at this interface between economic activity and a sustainable environment.

The environment is characterised in a number of ways. It is more commonly seen, at least by the public, to include protection of wildlife. When the world's nations signed up to the concept of sustainable development at the 'Earth Summit' in Rio in 1992, it was accompanied by additional treaties on maintaining *biodiversity* and on forests. However, the environment includes the protection of the human environment and maintaining the cultural heritage of communities.

Sustainable development cannot be achieved without economic growth. While some may question how economic growth needs to proceed in developed countries, the extensive poverty in the developing world cannot be eradicated without economic improvement. This must be achieved following the principles of sustainable development, otherwise the implications for the environments of developing countries and the rest of the world are bleak. The issue of environmental pollution is a key element in assessing the impacts of development on the environment, and managing this problem is a necessary prerequisite for making development more sustainable.

A number of key principles were identified in the Brundtland Report as necessary in order to achieve sustainable development. All of these have great significance for the management of environmental pollution and are worth examining in more detail.

Carrying capacity

The idea of 'carrying capacity' or 'environmental capacity' or 'environmental limits' has a wide acceptance in discussions on sustainable development, and much of this derived from earlier experiences with pollution control. The concept implies that there is a limit for the environment to absorb some form of activity, e.g. a population of people, number of cars, etc. For pollution it implies that there is a limit for the environment to absorb pollution.

The limit of pollution capacity can be considered in different ways. Some pollutants may be harmless in low concentrations and require a threshold for effect. Acid deposition, for example, is neutralised by base cations present in soils. A soil that can release very high levels of base cations (e.g. limestone or chalk) will have an almost unlimited ability to deal with incoming acidity. It has a very high carrying capacity. Rocks such as granite, however, do not. This is the basis of the 'critical loads' approach described in Chapter 3.

The environment might accumulate pollutants to a point where effects occur, i.e. there is a threshold determined by dose, not concentration. An example is an artificial wetland created to remove pollutants from contaminated water. Over time the accumulation of pollutants may mean that these begin to be leached from the wetland and it no longer performs its function.

For those pollutants without a threshold for effect, one management response is the 'dilute and disperse' method of control. By mixing the pollutants with a large quantity of one environmental medium, its effects become negligible. This was the philosophy behind the tall-stack policy for sulphur dioxide emissions in the UK in the 1950s and for discharging sewage to the sea. In both cases it proved erroneous. In small quantities the environment can cope with such pollutants, but the identification of a particular 'carrying capacity' is arbitrary.

The management of environmental risks: using the precautionary principle

Risk management is a key element of the work of a pollution manager. If substances are released into the environment, then a judgement has to be made of the risks that they pose and whether these risks are acceptable. This judgement can sometimes be based on sufficient scientific data to assess the risk, but more often there is an element of uncertainty, and for this one should follow the precautionary principle. This is described in the Brundtland Report:

> Risk assessment is one of the great challenges in sustainable development policy; the best available science is required to identify the hazards and their potential consequences, and to weigh up the degree of uncertainty. Where appropriate (for example, where there is uncertainty combined with the possibility of irreversible loss of valued resources), actions should be based on the precautionary principle if the balance of likely costs and benefits justifies it. Even then the action taken and the costs incurred should be in proportion to the risk. Action justified on the basis of the precautionary principle can be thought of as an insurance premium that everyone pays to protect something of value.

One reaction to uncertainty is to ignore anything that cannot be 'proven'. The precautionary principle acts as a moral injunction not to disregard such concerns. However, it has to be applied in a practical manner that does not lead to a prevention of development of any kind. It is possible to distinguish two practical forms of precautionary action. The first is *strict precaution*. This means that one would seek to prevent any development or pollutant discharge from occurring, and this would be appropriate where the uncertainty is combined with high potential damage. The second form is *adaptive precaution*. This acknowledges the uncertainty and, in allowing a development or discharge to proceed, builds into it procedures, monitoring, etc., that can assess whether adverse changes occur, and thus has the flexibility to respond to these.

There are many reasons why uncertainty exists in dealing with the management of environmental pollution. These include:

1   Lack of time to collect sufficient information. Very often decisions need to be taken and it simply may not be possible to obtain the necessary data to reduce sufficiently the uncertainty. This is true especially when examining longer-term impacts, e.g. chronic health problems, effects of low-level radiation or subtle changes to ecosystems.

2   Some environmental systems are so highly complex that our ability to predict their responses is very limited.

3   Some aspects of the environment are still poorly understood. This is particularly so for the maritime environment.

4   It may be possible to undertake experimental studies to obtain sufficient

information. However, this might mean manipulating the very aspects of the environment that one is concerned to protect. This defeats the purpose of pollution regulation.

5   There may be significant practical problems in obtaining data. For example, one may need to deal with remote sites, deep seas or subsections of the human population that are difficult to identify.

6   The costs of obtaining sufficient information may outweigh the economic benefits of discharging the pollutants.

It is always possible to identify uncertainty – dealing with it is another matter. UK government guidance (DoE, 1995) argues, for example, that the precautionary principle should not be used to generate completely hypothetical impacts. Some understanding must exist of how a pollutant might cause an effect, although uncertainties may remain as to its magnitude, etc. It is, of course, possible that a novel pollutant could cause quite unforeseen consequences. For example, when chlorofluorocarbons were first released, they were seen as inert and impacts on the ozone layer were not predicted (although a theoretical prediction of such an impact was subsequently made before the ozone 'hole' was discovered). However, generally one ought to be able to identify a reasonable causative link between pollutant and potential impact.

The guidelines from the UK Department of the Environment (DoE, 1995) outlines five stages in assessing risks:

1   *Description of intention.* This requires a detailed identification of the nature of the receiving environment and of all the proposed releases to the environment, including spatial distribution and changes over time.

2   *Hazard identification.* This part of the assessment seeks to identify the properties of released substances that could lead to adverse effects on the environment, e.g. relative toxicity data or potential for retention in the environment. Hazard identification should also include an assessment of potential problems from unintentional operations of a process (e.g. an accident).

3   *Identification of the consequences.* This requires an assessment of the exposure of components of the receiving environment to released substances. For example, is dispersion in the environment likely to lead to sensitive organisms being exposed? Assessment of routes of action (e.g. absorption into the body via skin, lungs or ingestion) form an important part of this stage.

4   *Estimation of the magnitude of the consequences.* Having identified that sensitive elements of the receiving environment might be exposed to released substances, the magnitude of any effect should be assessed. Toxicological and other data are necessary for this stage, as well as an understanding of the behaviour of the receptors (e.g. the population dynamics of a sensitive species). The DoE (1995) defines effects on the living environment (other than humans) according to four categories.

Severe: a significant change in the number of one or more species, including beneficial and endangered species, over a short or long term. This might be a reduction or complete eradication of a species, which for some organisms could lead to a negative effect on the functioning of the particular ecosystem and/or other connected ecosystems.

Moderate: a significant change in population densities, but not a change which resulted in total eradication of a species or had any effect on endangered or beneficial species.

Mild: some change in population densities, but without total eradication of other organisms and no negative effects on ecosystem function.

Negligible: no significant changes on any of the populations in the environment or in any ecosystem functions.

For the non-living environment the categories are defined as:

Severe: effects might be irreparable damage to geological features.

Moderate: effects might be damage to structures which are present in limited numbers (such as Grade II listed buildings).

Mild: effects might be damage to commonplace present day structures which could be repaired.

Negligible: effects might be very slight damage to such structures.

5   *Estimation of probability of consequences.* This assessment requires the examination of a range of issues. The likelihood of equipment failure may be predictable due to past experience. However, the behaviour of a pollutant in the environment may not be quantifiable. The recommendation is that when quantifiable risks are not possible, the estimation should be given in ranges, e.g. by orders of magnitude or likelihood over a period of, say, 100 years. However, to do this does require at least some appreciation of a semi-quantitative estimate.

Routes of action for pollutants

A major part of risk assessment methodology is to define the routes and mode of action of pollutants for important endpoints in the environment. For an ecosystem, for example, it is necessary to decide which are the important species in the system and which are the critical ecological processes there that could be disrupted. It is then necessary to consider the different routes by which pollutants can reach these endpoints.

Many pollutants may have complicated routes of action. Fluoride, for example, in the atmosphere, has concentration thresholds for direct toxic effects on man, animals and plants. However, plants and animals can also be affected by the longer-term accumulation of fluoride in the soil, deposited from relatively low atmospheric concentrations. Thus the assessment of different routes of action may lead to the setting of standards that have a different character to those set for human health.

Obviously, the environment presents a range of different potential endpoints. The assessment procedure, therefore, requires an iterative process. A procedure might follow the following route:

1   Dispersion modelling could be used to identify the different routes by which the pollutant may be distributed in the environment.

2   Each potential endpoint can be examined for receptors that may be particularly at risk, e.g. children, sensitive species or critical ecological processes, and how important each of these components are.

3   More detailed modelling is then undertaken for particular critical endpoints, to define more clearly the routes of dispersion and the effects of possible control or other mitigating measures.

Setting standards

A key tool for the pollution manager is the use of environmental standards. These can be used to judge the status of a component of the environment following receipt of monitoring information, or to assess more readily the consequences of proposed or existing pollutant discharges.

Throughout this volume, extensive reference will be made to a wide range of standards, set for very different objectives. In examining the way that standards have been developed over recent decades, it is obvious that different approaches have been undertaken and it is important for a pollution manager to realise the nature, and hence limitations, of any standards that are being used.

Many standards, for example, are set to protect only a subset of the environment. Most commonly this would be human health, but there are also a range of other standards for the natural environment, etc. It is important that a pollution manager, in both regulatory action and communication with business, government and the public, assesses whether meeting a standard adequately protects the environment as a whole. For example, many air-quality standards under debate at present are set only for protection of human health or a freshwater standard may be set to protect a particular type of fishery and not the complete river ecosystem.

Standards are also developed in different ways, and their use should reflect this. The simplest type of development of standards is to examine data for thresholds for significant effects and then use this number as the standard. Good examples of this include critical levels of air pollution for vegetation (see Chapter 3) or phosphate standards for freshwater ecosystems (see Chapter 4). In toxicology, such as threshold is commonly referred to as the no effect level (NOEL), either measured directly or estimated by a statistical analysis of response data to pollutant exposure.

For many substances there is no threshold for effect. In this case a risk assessment is undertaken and a value is derived which is considered 'reasonable'. For example, the air quality standard for particulates for human health is not set at a level below which it is thought that there is no effect, but at a level below which the effects are thought to be minimal.

In other instances, standards are derived from well-defined toxicity test procedures. Commonly these may determine the concentrations of substances that cause a given proportion of the test population to die (e.g. LD_{50} – lethal dose causing 50 per cent mortality). These tests may use particular organisms (e.g. trout or *Daphnia*), so that comparative data across a range of substances are obtainable. However, these organisms may not be particularly sensitive to the pollutant being tested for. In this case it is common to incorporate a safety factor into the results obtained, e.g. dividing the LD_{50} result by 10 or even 100. In this case the standard set cannot be said to be a threshold for any known effect; it is a level considered to be 'safe'.

There is nothing 'wrong' in having standards set in a wide range of different ways – each is appropriate for its own use. However, it is very important for a pollution manager to understand the nature and, therefore, the limitations on the uses different standards. This is especially so with integrated systems of pollution control, where impacts on different media occur and where different types of environmental standards are being assessed.

Assessing impacts on different species

Harm to human health is judged by well-established risk assessment techniques. Its aim is to prevent death or illness of individuals and, in some cases, offence to human senses as well. Extrapolation of toxicity data on non-human species to produce overall criteria for environmental protection has many difficulties. It is evident that different species exhibit different sensitivities to different pollutants. To examine the threshold effects of a pollutant on all species in an ecosystem one might, for example, produce a normal distribution of threshold concentration against the number of species for which that concentration is appropriate. If this model is correct, and if sufficient species responses are understood, it is possible to predict the responses of all species by assuming that the known data are distributed within the normal distribution.

Such analyses assume that the data which are collected are obtained for a random selection of species within the study group and are, therefore, randomly distributed throughout the normal distribution. However, it is quite possible that studies could focus on more sensitive species. This could be for reasons of importance to the environment, or alternatively, due to experience in the use of these species in the past. In any case, by gathering data from this source, it would be inappropriate to place the data into the normal distribution curve (or any other statistical model). The data would obviously skew the model.

Thus, small collections of data on species response are very difficult to fit into a model as it is not possible to know how they are distributed with respect to responses for all other species. Without good evidence, assumptions as to how they do fit could produce very erroneous conclusions.

For the natural environment, the levels of protection of every individual are rarely appropriate. One instance where individual protection is necessary, is in the case of very rare species in small populations. Here each individual should be protected. However, most of the natural environment does not fit this category.

There are a number of problems in deciding what is to be protected in the natural environment. The aims of the UK Environmental Protection Act 1990 are to prevent 'harm' to ecosystems, but how should this be defined? An ecosystem is the sum of its components, but is it necessary to protect all components, or just those considered to be 'important'?

In most instances, prevention of harm to ecosystems means preventing the deterioration of populations within those ecosystems. Thus, the effects on a few individuals of a population with high fecundity probably would not be important (see, for example, the effects of the Chernobyl disaster in Chapter 6). However, this in part depends on how pollutant impacts on a few individuals may occur. There may be two mechanisms:

1   A few individuals in the population may be particularly sensitive to the pollutant and thus are killed at very low concentrations. If these individuals are otherwise healthy, then this increased sensitivity may be genetic. Loss of these individuals may, therefore, affect the genetic structure of the population and this itself may be undesirable and, therefore, classed as 'harmful'.

2   Individuals may be affected due to the fluctuating or other behaviour of the pollutant in the environment. Thus a standard set as an annual mean concentration may provide general protection, but a few peak concentrations may affect random individuals in some locations. In this instance, the effects may not be 'harmful', as no particular section of the population is being selectively affected and the overall genetic character remains intact.

A further problem arises in defining the period of exposure upon which effects are determined. For persistent toxins and carcinogens the potential for accumulation in an ecosystem is important. This can either be in the physical environment (e.g. the soil) or biological (e.g. bioaccumulation of fat-soluble toxins). The latter route may mean that species not directly exposed to the pollutant may, nevertheless, be affected by it. Simple measures of exposure based on laboratory studies may underestimate long-term and indirect effects and therefore care needs to be taken when considering the behaviour of the pollutant in the ecosystems concerned.

Polluter pays and user pays

The discharge of pollutants into the environment is probably the best example of how industrial and other activity has taken a free ride on the environment in its pursuit of economic gain. For many years, companies, farmers, etc., could discharge their waste without concern for the damage they caused or the costs they forced on others (e.g. for clean-up or health impacts).

A key principle in sustainable development is that these external costs of economic activity should be internalised. A car driver, for example, should pay the full economic cost of the environmental damage caused by the use of the vehicle. For industry the additional costs will be passed on to consumers, so in this case it is considered that the 'user' pays.

The aim is that by internalising these external pollutant (and other environmental) costs, the free market should ensure that less-polluting activities are not unfairly discriminated against. Currently, for example, the electricity market in the UK does not distinguish electricity produced from cleaner sources, such as a coal-fired power station fitted with flue-gas desulphurisation (see Chapter 3) from one without clean technology. In this case being 'environmentally friendly' can be financially disadvantageous, and this problem needs to be addressed. In some instances, the polluter pays principle may interact with non-market systems. A good example of this is agriculture within the European Union. Most common crops are heavily subsidised, so a free market is non-existent. Public money is, therefore, being used to create pollution. Internalising costs may aid reducing pollution, but reassessing the basis for subsidies to maximise financial *and* environmental benefits in combination may be a better route forward in this instance.

It is important to note that the polluter pays principle does not imply that anyone can pollute if they can pay. The principle, and the market benefits that result, have to be used alongside a system of environmental standards or other measures that may limit polluting activity altogether. Thus it may be necessary to increase the cost of motoring to discourage car use, or encourage the use of more fuel-efficient vehicles. However, it also may be necessary to close some streets during adverse weather conditions, to prevent smog formation, no matter how much a few car drivers may be willing to pay.

CASE STUDY: PAYING FOR WATER POLLUTION IN THE EUROPEAN UNION

Leek and de Savornin Lohman (1996) examined the use of charging systems for water-pollution control in all EU member states. They noted that all member states operated a charging system to some degree, but that the extent and purpose of the systems in place were quite diverse. All countries, for example, operate user charges for water treatment works. In some (e.g. Denmark, the Netherlands and the UK) the aim of the charging is to cover all of the construction and operating costs. However, for many countries (e.g. Greece, Portugal and Spain) only operational costs have been recovered. The costs of construction have been met from public funds. The study noted that, generally, those countries recovering more costs had a higher percentage of the population connected to wastewater treatment facilities.

The study found that incentive charges to large polluters were levied in only a few EU member states. The UK was unique in that polluters are charged only to cover the costs of administering integrated pollution control, i.e. each processor pays a standard fee, irrespective of the quantities of pollutants emitted. Apparently, this scheme does not aim to affect the quantities of pollutants discharged. Leek and de Savornin Lohman looked at the charging schemes of the other member states using such schemes, by examining two hypothetical discharges, one with only basic treatment and one that had undergone extensive removal of pollutants. Table 1.1 presents the results of this study. It can be seen that the overall charges vary considerably and that the financial incentives to reduce emissions also vary. The authors note work demonstrating that these incentives have already achieved results in Germany and the Netherlands. However, the size of the incentives, for example, in France would be unlikely to have much effect. In the latter country, reductions have been achieved by using money raised by charges to finance subsidies for pollution reduction measures.

TABLE 1.1   Costs of discharges (in Ecu in 1992) for five EU member states for two types of discharge, and the difference in costs between the two

Country	Charge for Discharge One (basic treatment)	Charge for Discharge Two (extensive treatment)	Difference in charges
Belgium (Flanders)	59,647	27,582	32,065
Belgium (Wallonia)	28,089	14,386	13,703
France (Rhine–Meuse)	4,241	1,316	2,925
Germany	52,132	4,458	47,674
The Netherlands	48,103	15,500	32,603
Spain	9,563	4,142	5,421

Source: Leek and de Savornin Lohman (1996)

`Leek and de Savornin Lohman recommend the wider use of both user and incentive charges for pollution control within the EU. They do recognise the need for a political will to achieve this at national and local levels. However, as more EU directives require that emission and environmental standards are achieved (and at the lowest cost), their more widespread adoption is likely to occur.

Economic analysis

Estimating the costs and benefits of action to manage environmental pollution is a key element in any decision-making process.

In assessing the use of environmental standards, it is not only important to understand their operation as 'threshold' concentrations for environmental impact: the 'prevention of harm' in the UK Environmental Protection Act 1990 is tempered with the requirement to use the 'best available techniques not entailing excessive cost' (BATNEEC, see Chapter 3) to achieve this. An understanding of environmental benefits and economic costs is therefore essential. If a standard is exceeded, this might be allowable to the regulator if excessive costs are necessary to achieve the standard. However, once a standard is exceeded, the amount of environmental damage will be very different for the different pollutants. Some pollutants have steep response curves, while others do not. Therefore, it is not only necessary to set standards, but also to understand the response above that level. This will ensure a more accurate assessment of the environmental costs of breaching a standard.

CASE STUDY: ASSESSING THE ECONOMIC IMPACTS OF AIR POLLUTION FROM TRANSPORT ON HUMAN HEALTH

Maddison *et al.* (1996) describe a methodology to assess the external costs of air pollution, particularly in relation to human health. The first stage in the assessment is to calculate the actual damage to human health due to changes in air pollutant concentration. This is found using:

$$\Delta H_{ij} = b_{ij} * POP_j * \Delta * A_{jt}$$

where Δ = 'change in'; H_i = health impact i per year; j = concentration of pollutant j emitted; t = fuel type (e.g. diesel, petrol, unleaded); b_{ij} = slope of the dose response function relating the health effect (i) to pollutant concentration (j); POP = population exposed; A_{jt} = ambient concentration of pollutant attributable to emission from specific fuel types.

To place an economic value on these impacts, a price (P_i) is incorporated:

$$P_i * \Delta H_{ij} = P_i * b_{ij} * POP_j * \Delta * A_{jt}$$

This produces an estimate of the external costs, for health impacts, for one pollutant type. The economic value can be produced by tools such as assessment of '*willingness-to-pay*' as well as calculations of economic losses due to ill-health from air pollution. For emissions from motor vehicles, for example, impacts from all pollutants emitted (including *secondary pollutants* such as ozone) would need to be separately calculated and the whole aggregated to produce an estimate of the full external costs.

There are, of course, problems with any such analysis. In particular this analysis relies on accurate information on pollutant emission and dispersion, a clear understanding of the relationship between pollutant concentration and the health impacts, and a robust analysis of the economic values. Problems relating to the first two are dealt with in the later chapters on air pollution. Many arguments exist among economists over the third area, i.e. the economic valuation of environmental impacts – in this case human health. In this study a 'willingness to pay' valuation has been made, i.e. an estimate is made of what financial costs people would be prepared to forgo in order to achieve a given environmental objective. Maddison *et al.* (1996) conclude that a 'value of a statistical life' is currently about £2 million (although some would challenge this number). Various estimates can be made of health effects other than death and Table 1.2 contains some examples of these.

By putting all of the data together on health impacts and economic valuation for each pollutant emitted by road transport in the UK, an overall estimate can be made of the economic costs of these emissions. Table 1.3 outlines these results.

TABLE 1.2 Examples of economic values assigned by different workers to health conditions

Health impact	Mean value (£)
Respiratory hospital admissions	10,200
Emergency room visits	390
Asthma	25
Chronic bronchitis	15,300
Hypertension	220
Coronary heart disease	57,520
IQ points loss per child	340

Source: Maddison *et al.* (1996)

The data in Table 1.3 can also be assessed by fuel type, as in Table 1.4

It can be concluded that significant external economic costs exist from the emissions from road vehicles in the UK. It is important to note that this case study only examined the effects on health; including the effects on natural ecosystems, buildings, etc., would add to the costs. It also shows that not all motor vehicles cause the same impacts. This is explored later in Chapters 2 and 3, but this point is important with reference to any attempt to transfer these costs directly on to the users of motor vehicles, such as by increases in fuel taxation. Taxation can be used to reduce

TABLE 1.3 Estimated costs to human health of road transport in 1993 in the UK

Pollutant	Effect	Number of premature mortalities per year	Total external health costs (£ millions)
Direct PM_{10}	Mortality	1,725	3,450
	Morbidity		2,100
SO_x (incl. indirect PM_{10})	Mortality	1,880	3,760
NO_x (incl. indirect PM_{10})	Mortality	2,000	3,990
	Morbidity		2,160
VOCs	Mortality	1,010	2,020
	Morbidity		850
Lead	Mortality	20	40
	Morbidity		240
Benzene	Mortality	30	70
Total			**19,720**

Source: Maddison *et al.* (1996)

TABLE 1.4 The external cost of road transport in the UK, by fuel type

Fuel type	Total external cost (£ millions)	Fuel sales (millions of litres)	Marginal external cost per litre (pence)
Diesel	11,765	14,000	84
Petrol	6,375	15,000	43
Unleaded	1,569	17,000	9

Source: Maddison *et al.* 1996

fuel consumption generally. For example, Goodwin (1992) found that a 10 per cent increase in fuel costs results in a 7–8 per cent reduction in demand. Taxes can also be imposed to target the most polluting fuels, and the differential taxation of leaded and unleaded fuel in the UK has been a significant means of achieving the rapid adoption of its use (see Chapter 3).

Monitoring

There is now a huge number of different monitoring techniques and strategies open to the pollution manager. Monitoring is an essential tool, which can be used to address a wide range of questions that may need to be answered. For example:

1   What are ambient or background pollution levels?
2   Are modelling predictions of pollution from a process met in reality?
3   How do pollutants behave in the environment (e.g. reactions with other pollutants, mobilisation from sediments, etc.)?
4   What is the response of the environment to receiving pollution (e.g. the reaction of people, plants or animals)?
5   What information does one need to persuade the public that a process is safe?

Later in this volume the issue of monitoring will be raised a number of times. For example, Chapter 3 will examine the relative value of simple and complicated air pollution monitoring techniques and Chapter 4 will examine the issue of biomonitoring in freshwaters. Monitoring can, of course, be undertaken at a range of scales. It is necessary to monitor air pollution around a power station, or water chemistry in a river receiving effluent. It is also important to undertake regional monitoring, for example to assess the changes in general air quality in a city. Many policy decisions are nationally based, and country-wide monitoring networks are essential to inform future decisions. Finally, of course, international cooperation on monitoring is essential, as much pollution crosses national frontiers, e.g. monitoring acid rain across Europe, the transfer of pollutants in marine waters or the movement of radionuclides from the Chernobyl accident. International cooperation in the European Union was enhanced by the recent formation of the European Environment Agency (EAA) based in Copenhagen. Currently, the work of the EAA has focused on establishing 'topic centres' in each member state to coordinate the supply of environmental monitoring data to produce a clearer picture of the state of the environment within the EU and how this might be used to aid production of future EU legislation.

Monitoring can be perceived as a very 'busy' activity for a pollution manager producing lots of 'results'. However, it is vital that the effort given to monitoring is properly targeted, otherwise the data collected will have limited value. Collecting data is no substitute for clear analytical thinking. It is perfectly possible to be 'data rich and information poor'. There are now a vast number of pollution monitoring networks and sites in, for example, the UK. Even with these, there are still some basic questions that cannot be answered adequately. It is important to target monitoring at meeting specific objectives or testing particular hypotheses. If monitoring is being undertaken to assess water or air quality at a particular location linked to a specific process, it is important that specific action is linked to the results. A high pollutant level might, for example, lead to the process being shut down. Much monitoring is linked to the concept of 'limits of acceptable change', i.e. possible fluctuations in pollutant levels or the numbers of a particular species in the environment, but there is a point where a management response is deemed necessary. Nitrogen dioxide levels in the atmosphere may change within a city, but if they reach a level deemed 'unsafe', then it may be appropriate to prevent further car use. An occasional instance of imposex in dogwhelks in coastal waters may not be a problem, but if a significant proportion of the population begin to exhibit the condition, a re-examination of the use and behaviour of tributyltin in the area would be needed.

One of the biggest obstacles to good information from monitoring is that a wide range of organisations undertake monitoring, although not always in relation to each other. It is sometimes difficult, therefore, to compare information and thus form a more holistic view of the environment. The Environment Agency (EA) in the UK is currently attempting to rectify this by proposing a collaborative forum, bringing together the major organisations that undertake environmental monitoring to agree a common framework and overcome problems of data sharing, etc. They currently consider that the framework could have six components:

1   The use of land and environmental resources.
2   The status of biological communities and populations.
3   The chemical quality of the environment with respect to existing standards and targets.
4   The 'health' of environmental resources (e.g. biomonitoring).
5   Environmental changes at long-term reference sites.
6   The aesthetic quality of the environment.

Ultimately the aim of this monitoring framework would be to aid regulatory and land-use management decisions, and produce better informed policy and legislation.

Chapter summary

- There are a large number of basic principles and concepts that a pollution manager has to consider. Even defining 'pollution' can be problematic, especially as it may have contrasting inferences in different cultures. This book takes a broad definition of pollution, which includes natural as well as man-made substances or energy (e.g. heat) that may have an adverse impact on human health or well-being or on the natural or cultural heritage.

- Pollution management has to be set in the context of achieving sustainable development, i.e. integrating the objectives of environmental protection, social needs and economic development. There are a number of concepts central to sustainable development:

- *Carrying capacity.* This is the ability of the environment to absorb activities such as pollution without adverse impacts. This is the basis for such concepts as 'critical loads'.

- *Precautionary principle.* Our ability to understand the environment is limited and it should not be necessary for every aspect to be understood prior to any action to reduce pollution. Detailed procedures are now available to assess the management of uncertainty.

- *Polluter pays and user pays.* Historically and currently, many polluters have had a 'free-ride' on the environment, treating it as a disposal point without any cost to them. Consumers have not paid the true costs of their consumption either. This principle aims to ensure that such external costs are fully internalised in the economic activities. The first case study examines how charging for water pollution in different EU countries may lead to incentives for water treatment. The second case study examines attempts to describe the economic consequences of road transport pollution on human health.

- Pollutants may act in different ways in different media and on different species. The different routes of dispersion and types of impact are described, e.g. distinguishing effects on individuals and on populations. The role of standards is very important. These are set for very different purposes, some for nil effect, some for 'acceptable' impacts, for human health or the natural environment. A pollution manager must understand the nature and, therefore, limitations of any standards that are being used.

- The role of monitoring in pollution management is important. In all cases it is necessary to target monitoring at specific objectives, whether these be on the operation of a specific industrial site or the assessment of the success of pollution control legislation at a national or international level. The European Environment Agency is collating a large amount of monitoring data across the European Union, and the Environment Agency in the UK is proposing a collaborative forum to produce an overall environmental monitoring framework for the UK.

Recommended reading

DoE (1995). *A Guide to Risk Assessment and Risk Management for Environmental Protection*. HMSO, London. This is an excellent guide, taking the reader carefully, step by step, through the operation of the precautionary principle. It examines the problems of obtaining sufficient information to make judgements in environmental management and attempts to offer a reasonable solution to dealing with these problems. While the book deals with all aspects of environmental management, the control of pollution is frequently discussed.

Maddison, D., Pearce, D., Johansson, O., Calthorp, E., Litman, T. and Verhoef, E. (1996). *Blueprint 5: The True Costs of Road Transport*. Earthscan, London. This is one of the best case studies of environmental economics. Road transport has enormous environmental effects, from land-take to health impacts. This book is an exhaustive study, which examines, in turn, the greenhouse effect, air pollution, noise, congestion, accidents and aggregate use. It is set in a UK context, but there are case studies from Sweden, the Netherlands and North America. Road users do not pay for the consequences of their actions, and the book demonstrates this admirably.

WCED (1987). *Our Common Future (The Brundtland Report). Report of the World Commission on Environment and Development*. Oxford University Press, Oxford. This book is the argument for sustainable development, which led to the Rio conventions in 1992. Global problems are well summarised and the developed world is challenged to change its approach to economic development if we are to protect the environment and meet the social needs of the developing world. It is a powerful statement. Many people refer to the Brundtland report, but a good proportion of them have not read it. The consequences of this are that they take one or two definitions from the book and argue over their meaning, whereas if we read the volume itself their meaning would be clearer from the context in which they are used.

Chapter 2

Air pollution
and its impacts

◆ Introduction   22
◆ Air pollutants and their sources   22
◆ Air pollution impacts on health   34
◆ Impacts of air pollution on the natural
  environment   40
◆ Impacts of air pollution on the built
  environment   51
◆ Impacts on visibility   53
◆ Economic impacts   54
◆ Chapter summary   55
◆ Recommended reading   56

Introduction

The range and sources of air pollutants are almost limitless – as are the public's concerns about them. It is only necessary to examine press reports over the past two decades to see how concern over different air pollutants has risen in the public view and then declined. There have been the large global and regional concerns such as those of *acid rain*. However, even this has changed in character. Initially, concern was voiced on sulphur dioxide emissions from power stations and the acidification effects that these were having. However, this soon included concern over nitrogen oxide emissions from power stations and motor vehicles. The deposition of nitrogen also adds nutrients to ecosystems, so attention has also turned to the emissions of ammonia from agriculture. Nitrogen oxides also contribute to ozone formation in the lower atmosphere when they react with volatile organic compounds. Thus, quite quickly, the 'simple' story is complicated by a wide range of pollutants with many sources and impacts.

The impacts on health have been equally varied. In the beginning, serious concern surrounded local sulphur dioxide emissions and particulate sources, predominantly domestic ones. However, as these began to be controlled, concern centred on particles and nitrogen oxides from motor vehicles.

The key lesson for the pollution manager seems to be that new air pollution problems arise as quickly as old problems are solved. While the above examples are of national and global concern, the pollution manager should also recognise that *local* air pollution issues can also be serious. Individual sources of toxic emissions may threaten local communities (see later case study on dioxin emissions), or nuisance may be caused by dust from a quarry.

This chapter will initially examine some of the sources of a few key air pollutants and the impacts that they have. The pollutants considered include ammonia, carbon monoxide, fluorides, heavy metals, nitrogen oxides, ozone, particulates, sulphur dioxide, volatile organic compounds and acid rain. A detailed account is then given of the effects that these may have on human health, on the natural environment, on the built environment, on visibility and the economic consequences. Chapter 3 will address the issues of air pollution regulation, monitoring and management.

Air pollutants and their sources

Surveying emission sources can either be done on the basis of individual pollutants or individual sources. Motor vehicles, for example, emit a cocktail of nitrogen oxides, carbon monoxide, volatile organic compounds, particulates, heavy metals and (for

BOX 2.1 Units of measurement

Most pollutants are measured using just one method (although older studies may refer to systems of measurement that are no obsolete). However, for air pollutants it is common to express concentration in two different ways. This may either be the mass of a pollutant per unit volume of air (e.g. mg/m³ or μg/m³) or the volume of a pollutant per unit volume of air (e.g. parts per million (ppm), or parts per billion (ppb)). For many purposes, comparisons between pollutants are easier on a ppm or ppb basis, i.e. air containing 10 ppm of sulphur dioxide contains the same number of pollutant molecules as 10 ppm of ozone. It is more common, however, to see the use of mass per unit volume, and emission standards for industrial plant are always expressed in this fashion. Occasionally both forms may be found. For example, some pollutants in the air quality standards in the UK National Air Quality Strategy (see Chapter 3) are expressed on a mass basis and some on a volume basis. The conversion between the two measurement system will vary depending on temperature and pressure. A few example conversions are given in the table below for atmospheric pressure and 20°C. These conversions would not be appropriate, for example, in the high temperatures of gases in chimney discharges. The conversions are simple, thus 1 ppm ammonia is equal to 0.71 mg/m³.

Pollutant	*1 ppm = x mg/m³* *1 ppb = x μg/m³*	*1 mg/m³ = x ppm* *1 μg/m³ = x ppb*
Ammonia	0.71	1.41
Hydrogen fluoride	0.83	1.20
Nitrogen dioxide	1.91	0.52
Ozone	2.00	0.50
Sulphur dioxide	2.87	0.38

It is also common to see a reference to 'percentiles' in air pollution monitoring information. For example, it may be stated that the annual mean sulphur dioxide concentration is 15 μg/m³, but the 98th percentile is 105 μg/m³. A percentile is a useful way of expressing peak concentrations. In the above example, 98 per cent of the monitoring samples recorded were below 105 μg/m³. This is better than providing information on the very highest peak measurement alone, as this might be an extremely anomalous reading. Air quality standards or industrial emission controls may be expressed both as mean concentrations and as percentiles, in order to control overall pollutant levels and peaks of pollutant exposure.

diesel) sulphur dioxide. Some regulation (e.g. on motor vehicle emissions) is aimed at sources. However, others (e.g. international agreements through the United Nations Economic Commission for Europe) are aimed at individual pollutants. This chapter will, therefore, survey some of the more important air pollutants on an individual basis.

Ammonia

Ammonia is produced from a few industrial processes, but by far the biggest source is agriculture. The intensification of animal production (e.g. pig or poultry units), and increasing use of fertilisers has caused a large increase in emissions in recent years. Emissions from animals occur from the unutilised nitrogen excreted in their waste. These emissions may occur as the waste is produced, or during its storage or spreading on fields. Emissions may also occur from fields where excess nitrogen fertilisers are applied. This latter problem is also a cause of nitrate ingress into groundwaters and is a serious water pollution problem (see Chapter 4).

It is difficult to assess accurately total ammonia emissions for a given area, as the emission rate varies considerably. Estimates for total UK emissions, for example, have ranged from 186 to 585 kt N/yr (CLAG, 1994a). Table 2.1 provides estimates of the percentage contribution to total emissions from different sources from two recent studies.

TABLE 2.1 Estimates of the percentage contribution to total UK annual ammonia emissions from two studies

Source	Study	
	Lee and Dollard (1994)	Sutton et al. (1994)
Cattle	51	55
Pigs	5	9
Sheep	17	9
Poultry	5	9
Fertiliser application/crops	10	8
Industry	2	2
Other animals	4	3
Other (sewage, landfill, etc.)	6	5

Carbon monoxide

Combustion of the carbon in fossil fuels leads to the production of large quantities of carbon dioxide (a greenhouse gas and therefore outside the scope of this volume).

However, not all such combustion is complete, and partial oxidation of the carbon can lead to the production of carbon monoxide, which is toxic to humans. This is particularly so where the combustion processes may be enclosed and have limited access to atmospheric oxygen. The internal combustion engine is a good example of this, and so most of the concern over carbon monoxide emissions is regarding motor traffic, which is responsible for 90 per cent of UK emissions. The major source on the roads is the private car (Table 2.2). As with other vehicle pollutants, emissions have increased rapidly in recent years, with a 30 per cent increase over the last decade. However, with the introduction of catalytic converters (see Chapter 3) the emission levels of many pollutants are predicted to decrease, although this will be partly offset by increasing traffic levels (Table 2.3).

TABLE 2.2 Percentage of aerial emissions of pollutants from motor vehicles, by vehicle type, in the UK in 1990

Pollutant	Cars	Light goods vehicles	Heavy goods vehicles	Other vehicles
Carbon monoxide	88	7	3	2
Nitrogen oxides	72	7	19	3
Volatile organic compounds	84	7	6	3
Particulates	6	7	77	10
Sulphur dioxide	37	9	47	6

Source: RCEP (1994)

TABLE 2.3 UK 1990 emissions of pollutants from motor vehicles and the range of projected emissions for the year 2025, high and low forecasts

Pollutant	1990 emissions (kt)	2025 emissions (kt)
Carbon monoxide	7,216	2,144–2,652
Nitrogen oxides	1,463	623–810
Volatile organic compounds	960	225–285
Particulates	45	17–22
Sulphur dioxide	63	41–53

Source: RCEP (1994)

Fluorides

Many minerals contain fluorides and, when used by industry, may result in the emission of fluorides into the air. The most common are gases such as hydrogen fluoride, but it is usual for other gases, aerosols and particulates containing fluoride to also be emitted. Hydrogen fluoride is a very reactive gas, so natural background levels are extremely low. It can, however, cause direct damage to some plants. Nevertheless, of most concern is the accumulation of fluoride in the soil, with its eventual toxic effects on plants, animals and, ultimately, humans.

The industrial processes best known to produce fluoride emissions are aluminium and uranium smelting, steel production, phosphate fertiliser production and brick manufacture. Some of these can produce sufficient hydrogen fluoride to cause extensive local impacts. However, emissions are not so great as to cause general national or international problems.

Heavy metals

A large number of metals are toxic to plants and animals. Many do occur naturally in varying concentrations, due to rock and soil weathering and volcanic activity. However, human activity can result in large additional emissions. Most metal processing operations will emit varying quantities of metals, as is to be expected. However, there are also other significant sources, such as fossil fuel burning (coal and oil contain traces of heavy metals). The most publicised heavy metals emissions are those of lead in motor vehicle exhausts, where the lead is specifically added to petrol as an anti-knocking agent. The metals of most concern are arsenic, cadmium, cobalt, copper, chromium, iron, lead, manganese, mercury, nickel, silver, tin, titanium, vanadium and zinc. Aerial emissions do constitute important sources and impacts of heavy metals. However, it is arguable that more extensive impacts do occur in freshwater and marine ecosystems, and therefore in this book, further consideration of the problem of heavy metals can be found in later chapters.

Nitrogen oxides

Nitrogen oxides are produced by the combustion of all fossil fuels, from coal- and gas-fired power stations to motor vehicles. Some of the nitrogen can come from the fuel itself. However, most of it comes from the reaction of atmospheric nitrogen and oxygen within the combustion chamber. As a result, it is not possible to control nitrogen oxides by changing fuel quality. Control is achieved either by affecting the combustion conditions (e.g. 'low-NO_x' burners for power stations) or by cleaning the exhaust gases (e.g. catalytic converters on motor cars).

There are two forms of nitrogen oxides – nitrogen monoxide (NO) and nitrogen dioxide (NO_2). Initially, most of the nitrogen oxides emitted from a particular process are in the form of monoxide. However, in the air this is oxidised to the dioxide. For convenience the two forms are referred to by the collective term 'NO_x'.

Table 2.4 shows the different UK sources of NO_x for 1991, and comparative figures for 1981. High levels of NO_x occur when sources are most active. Peak concentrations occur, therefore, during peak electricity demand and during high

TABLE 2.4   Estimated emissions of nitrogen oxides for different emission sources in the UK

Emission source	Emissions (kt) 1981	Emissions (kt) 1991
Domestic	68	76
Commercial	61	60
Power stations	839	718
Refineries	38	37
Agriculture	5	4
Other industry	268	224
Railways	39	31
Road transport	810	1,400
Civil aircraft	9	14
Shipping	104	133
Offshore oil and gas	56	50
Total	**2,297**	**2,747**

Source: PORG (1993)

PLATE 2.1   Modern smogs develop rapidly in urban areas, where there are mixtures of nitrogen oxides, volatile organic compounds and particulate emissions. This is especially so in hot, dry conditions, such as are beginning to occur here in Sydney.
Photograph: Andrew Farmer.

traffic volume. Thus, in urban areas, peaks in NO_x will occur during the morning and evening rush hours.

Ambient concentrations of nitrogen dioxide are much higher in urban than rural areas. For example, a typical busy road in London can show levels of well over 100 $\mu g/m^3$ as a daily average. During smogs, these concentrations can go much higher. For example, in the two worst recent London smogs of 1991 and 1994, peak concentrations were found to be 809 $\mu g/m^3$ and 551 $\mu g/m^3$ respectively as hourly averages. While rural nitrogen dioxide levels are much lower (in western Scotland they may be as low as 1 $\mu g/m^3$), it is now increasingly common to find rural levels above 30 $\mu g/m^3$ as an annual average in Western Europe (above the *critical level* for vegetation – see p. 66 (Ashenden and Edge, 1995).

Ozone and other photochemical oxidants

Ozone is unusual among air pollutants in that it does not have an emission source. It is a *secondary pollutant*, produced by the reaction of *primary pollutants*, nitrogen oxides and hydrocarbons, in the presence of sunlight. Because ozone is a strongly oxidising chemical and its production requires the presence of sunlight, it is known as a photochemical oxidant pollutant. Ozone does occur naturally: indeed the high levels in the stratosphere are necessary to protect the Earth's surface from excessive ultra-violet radiation. Low levels do occur in the lower troposphere, but the increasing concentrations resulting from the emissions of primary pollutants are of most concern for human health and have impacts on the natural environment. Indeed, such impacts may occur at concentrations not far above natural background levels, thus making the management of this pollutant more complicated.

Hydrocarbons are degraded in the atmosphere by the presence of hydroxyl *free radicals* (OH). This produces a peroxy radical (RO_2). The peroxy radical oxidises nitrogen monoxide (NO) to nitrogen dioxide (NO_2). The nitrogen dioxide may then dissociate to nitrogen monoxide, producing an oxygen free radical (O). This can then react with oxygen to produce ozone (O_3). Other secondary photochemical pollutants may be produced by further reactions. These include hydrogen peroxide and peroxyacetyl nitrate. Neither of these occur in concentrations as great as ozone. In areas where high nitrogen monoxide concentrations occur (e.g. urban areas) the additional NO may react with any ozone that is present to oxidise to NO_2. Thus ozone is generally considered to be a pollutant problem in rural areas, unless very high concentrations of hydrocarbons and high intensities of sunlight are present, as, for example, in Athens or Los Angeles.

The reactions producing ozone require a number of days to complete. Thus, the production of the pollutant may not occur near the source of primary pollutant emissions. For example, emissions of volatile organic compounds and nitrogen oxides may take place in mainland Europe, but the ozone may not be formed until the air passes over the UK.

As ozone production requires sunlight, peak concentrations tend to occur around midday and early afternoon in sunny conditions. At night the presence of nitrogen monoxide will deplete the ozone that has been formed. However, in upland areas it has been found (PORG, 1993) that this strong diurnal variation in ozone concentration is less marked, and that high concentrations can be maintained throughout the night.

Natural background levels of ozone are about 20–40 $\mu g/m^3$ as a summer average. However, over much of Western Europe, ambient levels are about twice this and are thought to be increasing at the rate of about 1–2 per cent per year (PORG, 1993). For many health impacts, peak concentrations are of most concern, and these often reach 160 $\mu g/m^3$ and occasionally over 200 $\mu g/m^3$ in Western Europe. However, in the world's most polluted cities, such as Mexico City, peak concentrations may be five to six times this level.

Particulates

The term 'particulates' covers a wide range of pollutants, ranging from visible dusts that may act as a neighbourhood nuisance, to very small particles that may cause considerable health problems.

Dusts may be defined as solid matter that is small enough to be raised and carried by the wind. A number of industrial sources produce dusts. Many of these can now be relatively easily controlled through the use of cyclones and electrostatic precipitators. Harder to control are sources from the mineral industry (e.g. quarrying) and from unpaved roads. These latter sources tend to have higher emission rates during periods of heavy vehicle use and in dry weather.

Of particular concern for human health are particulates with a diameter of 10 μm or less, the so-called PM_{10} (QUARG, 1996). These are considered in greater detail on p. 36. However, the total annual UK emissions of PM_{10} are estimated at 263,000 tonnes (EPAQS, 1995a). Table 2.5 provides a breakdown of the percentage contribution from the different sources.

The management of particulate pollution has one advantage over that of many other air pollutants, i.e. it is sometimes possible to identify the source of the pollutants by a study of the particles themselves. While, for example, an examination of sulphur isotopes may indicate the source of sulphate deposition (e.g. oil burning or maritime sources), an analysis of the particles may indicate not only the type of source, but the specific source. Particles are of most concern in urban environments and are produced by a range of activities, e.g. car exhausts, industrial emissions and construction. Recent research at the Massachusetts Institute of Technology has shown that the examination of particles under an electron microscope can identify the particles, as each pollution source has its own signature or 'sootprint'. Early analysis is being undertaken to determine the pollution sources in Boston, but this type of analysis is likely to be more widely used by pollution managers and policy-makers throughout the world.

It is also important to note that dusts can be generated naturally from disturbed soil. Areas of high winds with significant agricultural land can experience large dust clouds. In the UK the 'dust-blows' of the fenlands in East Anglia are a good

TABLE 2.5 The percentage contribution to PM_{10} emissions in the UK by different sources

Source	Percentage of total emissions contributed by the source
Power stations	15
Domestic	14
Commercial/public service	2
Refineries	3
Iron and steel	8
Other industrial combustion	7
Construction	2
Industrial processing	11
Mining and quarrying	11
Extraction and distribution of fossil fuels	0
Solvent use	0
Road transport: diesel	19
Road transport: petrol	5
Road transport: non-exhaust (tyres and brakes)	2
Other transport	3
Waste treatment and disposal	0
Agriculture	0
Total	**100**

Source: EPAQS (1995a)

example. Perhaps the most extreme case of this occurred in 1988 when soil was carried by wind from China to the Canadian Arctic. There it turned the snow brown over an area of about 20,000 km². An additional problem in this instance was that the soil itself was contaminated with pollutants such as polycyclic aromatic hydrocarbons and polychlorinated biphenyls.

Sulphur dioxide

Sulphur dioxide is produced by the combustion of fuels containing sulphur. It is rare for it to occur naturally, and then only close to some volcanic sources. The main combustion sources are generally coal and oil. Some oil-derived products lack sulphur (e.g. petrol), while others retain it (e.g. diesel). Historically, the extensive use of coal for domestic heating was a significant national and local source of sulphur dioxide (and particulates). However, this has now declined substantially, so that the principal sources are the large power stations and industrial processes that use coal or oil as energy sources. Natural gas is very low in sulphur, so its combustion is not a source of sulphur, and in the UK the current switch to gas-fired

TABLE 2.6 UK emissions of sulphur dioxide, by source

Source	1970		1993	
	Emission (kt)	Percentage of total	Emission (kt)	Percentage of total
Domestic	522	8	113	4
Iron and steel	435	7	92	3
Other industry	1,804	28	509	16
Power stations	2,913	45	2,089	66
Refineries	213	3	156	5
Road transport	44	1	59	2
Shipping	129	2	51	2
Other	368	6	119	3
Total	6,428	100	3,188	100

Source: EPAQS (1995b)

power generation has made a major contribution to achieving sulphur reduction targets. Peak concentrations of sulphur dioxide tend to occur during periods of high electricity demand. Table 2.6 demonstrates how emission sources in the UK have altered over the last two decades.

In much of Western Europe and North America, ambient concentrations have fallen markedly in recent years (Laxen and Thompson, 1987). Whereas it was possible to find winter mean concentrations in Britain of over 150 $\mu g/m^3$ in the 1950s, it is now rare to find areas where winter means are above 10 $\mu g/m^3$. In Europe the area with the worst sulphur dioxide concentrations is known as the Black Triangle, comprising the region where the Czech Republic, Germany and Poland meet. A long history of burning very-high-sulphur lignite coal has resulted in high annual concentrations and peaks in some cities of 1,000–2,000 $\mu g/m^3$. A similar story can be found in China, where high sulphur dioxide levels occur in coal producing regions. China adopted a strategy of locating lots of small industry in towns throughout the country, rather than concentrating particular types of production (e.g. steel-making) in a few large centres. This means that pollution control is very expensive and air pollution is still extensive, with smogs showing peak concentrations of around 900 $\mu g/m^3$.

Organic toxins

A number of organic compounds emitted into the atmosphere are particularly toxic. Well-known examples include polychlorinated biphenyls (PCBs) and polyaromatic hydrocarbons (PAHs). Emissions of these chemicals often lead to high-profile public concern. Sometimes this is without reasonable cause. However, well-publicised impacts such as the effects of Agent Orange in Vietnam, or the deaths and long-term illness produced by dioxin release from Seveso in 1976, demonstrate the impacts such pollutants can have.

PCBs are highly toxic and very persistent in the environment, and some of these are further considered in Chapter 5. PCBs have been used widely in plastics, hydraulic fluids and fire retardants for many decades. Disposal of materials containing PCBs has often led to environmental contamination. The focus today is on the destruction of materials by incineration. This has be undertaken at temperatures above 900°C, and with a residence time of greater than two seconds. PAHs, such as benzopyrene, are derived from motor vehicles, and some other combustion processes, such as those involving wood. PAHs have immunological and carcinogenic effects.

A number of these organic toxins, including many pesticides, can be found in increasing concentrations in the polar regions of the world, many thousands of miles from where they were used. For example, the Inuit people of Broughton Island in the Canadian Arctic have PCB concentrations in their body tissues comparable with those found in people exposed to industrial accidents. It is suggested that use of, or emissions of, such volatile organic chemicals in tropical regions leads to their evaporation into the atmosphere. They remain in the air until they reach much colder climates, where they condense. As the Arctic area is very much smaller than the tropics, this may lead to significant pollutant concentrations.

Volatile organic compounds

The term 'volatile organic compound' (VOC) is applied to a wide range of organic air pollutants, from simple hydrocarbons to more complex molecules. Some of these are produced during various production processes (e.g. emissions from road vehicles due to incomplete combustion) or are simply lost through the vaporisation of chemicals when used (e.g. solvents used in paint spraying or evaporation of petrol at pumping stations). The importance of different sources is shown in Table 2.7. Some of these compounds can have irritant or direct toxic effects. However, the main reason for grouping these substances together is because of their role in ozone formation (see p. 28). The degree to which individual VOCs contribute to ozone formation depends, of course, on the quantity emitted. However, each also has different reactivities in the atmosphere. Thus each VOC has been ascribed a 'photochemical ozone creation potential' (POCP), describing the degree to which it reacts to form ozone compared with an arbitrary 'standard' VOC, which is ethene. Thus controls to reduce VOC emissions in order to reduce ozone pollution should focus on those VOCs with the highest POCPs.

Wet deposited acidity or 'acid rain'

The term 'acid rain' is interpreted in various ways by different authors. Some use it to include all acidifying and oxidant air pollution, others limit it to wet deposited

TABLE 2.7 Estimated emissions of volatile organic compounds in the UK in 1990 and 1980

Emission source	Emissions (kt) 1980	Emissions (kt) 1990
Domestic	80	40
Commercial	1	1
Power stations	14	13
Other industry	41	41
Processes and solvents	1,160	1,189
Gas leakage	31	34
Forests	80	80
Railways	10	8
Road transport	864	970
Civil aircraft	3	4
Shipping	13	14
Total	**2,297**	**2,395**

Source: PORG (1993)

acidity. For the purposes of this volume, the term shall be limited to the latter, i.e. acidifying air pollutants deposited in rain, snow, cloud and mist.

The two principal primary causes of acid rain are sulphur dioxide and nitrogen dioxide. As these pollutants are transported over long distances, they are oxidised in the air to form sulphuric acid and nitric acid. Typically, the deposition of such pollutants may take place many hundreds of kilometres from their source. Indeed a small proportion of the acid rain deposited in north-west Europe is derived from North America. It is this international nature of the emission/deposition relationship with acid rain that is the reason for considerable diplomatic efforts by some countries to deal with the problem.

Once oxidised in the air, the sulphuric and nitric acids will dissolve in the available cloud water. Rain is already naturally acidic, due to small amounts of dissolved carbon dioxide. Unpolluted rain has a pH of about 5.7. Acid-rain pH varies greatly. Much of the rain in eastern England, for example, may be around pH 4.5. However, severe pollution events may cause rain or mist of pH 2.0 to be formed. The effects of acid rain often depend generally on the total quantity of acidity that is deposited. In eastern England, for example, the rain is more acid than in the west of Britain (being close to a number of emission sources). However, there is much higher rainfall in the west of Britain, so that the total deposited acidity is higher there. The less water available, the more concentrated the resulting acid solution becomes. Thus, while most public perception is focused on *acid rain*, it is the acid mists of many upland areas that may contain the highest acid concentrations. These may deposit directly on to foliage and other surfaces: this is known as '*occult deposition*'. The rain clouds above this mist may also provide rainfall through it, which picks up acidity as it passes through. This is known as the 'seeder–feeder' effect. Finally, snow may also contain high quantities of acidity. However, this is locked up in the snow beds until the spring thaw, so that all the available acidity is flushed into the surrounding soils and freshwaters in one large pulse.

Air pollution impacts on health

The impacts of air pollution on human health are of major concern to pollution managers. It is necessary to separate out both *acute* effects, such as those produced by pollution episodes, and longer-term *chronic* effects, which may affect a greater proportion of the population. Health effects are a very sensitive issue politically. It is generally true that the sections of the population most sensitive to most air pollutants are the young, the old and the sick. Ill-health in these groups also generally produces a greater emotional response among the population than do similar problems for young adults. A 7-year-old child using an inhaler to control asthma somehow seems more unfair than an adult using one – although, of course, both are innocent.

Acute effects are more easily distinguished. In the UK, the most notorious air pollution incident was the London smog of 1952 which, over the space of five days, caused an additional 3,500–4,000 more deaths over and above those expected for the period. It is curious that at the time no one really noticed what was going on. Today if there is a pollution incident known to be taking place, pictures of children walking to school alongside traffic-laden streets would be on the television. However, the numerous deaths in 1952 only became apparent when the statistics were collated shortly afterwards, and questions began to be asked.

More recently, concern over the effects of air quality in London has increased again. A severe winter pollution episode in December 1991 was judged to have killed between 100 and 180 people, and additional research has been undertaken since then to pinpoint the main pollutants responsible. Correlations were found between daily mortality and pollutant concentrations of particulates, nitrogen dioxide, ozone and sulphur dioxide. Both general mortality and that from respiratory and cardiovascular disease were examined separately. A positive correlation was found between mortality and ozone in the spring and summer and with particulates throughout the year. An increase in eight-hour ozone levels from 4 to 36 ppb, for example, resulted in an increase in general mortality of 3.5 per cent, in that from respiratory disease by 5.4 per cent, and in that from cardiovascular disease by 3.6 per cent. A 10 $\mu g/m^3$ increase in particulates produced an increase in mortality of 1.1 per cent. The authors conclude that 'it would be prudent to assume that current levels of air pollution do have adverse health effects'. The 1996 UK National Air Quality Strategy (see Chapter 3) estimates that, each year, air pollution causes several thousand deaths and up to 20,000 hospital admissions.

Impacts on health are also economically important. Those who become sick due to increased asthma or other diseases cannot perform productive work. There is also the direct cost of health-care treatment. For example, in the UK, there has been a rise in asthma of 67 per cent since 1990. In 1990 the costs of treatment were £245 million, while in 1995 this had risen to £410 million. There are, therefore, strong economic reasons for tackling air pollution problems affecting health. In the United States, recent studies have also shown dramatic health and economic benefits arising from the controls in the Clean Air Act. Over twenty years this has resulted in savings of $6 trillion. In 1990, for example, the controls resulted in a reduction of 79,000 deaths, 15 million cases of respiratory illness, 13 million cases of hypertension, 18,000 heart attacks and 10,000 strokes.

Carbon monoxide

Carbon monoxide is a poison that reduces the ability of the body to absorb oxygen. Oxygen is taken up by the haemoglobin in the bloodstream. However, carbon monoxide competitively interferes with this process by binding to the haemoglobin to form carboxyhaemoglobin. The affinity for haemoglobin of carbon monoxide is over 200 times stronger than for oxygen, so the ability of the blood to take up oxygen is quickly reduced. Foetal haemoglobin has an even greater affinity, so that carbon monoxide taken up by the mother will be specifically passed on to the foetus. Impairment of oxygen supply to the developing foetal brain is particularly dangerous, as it is at this stage that damage can occur very easily.

At low concentrations, carbon monoxide can decrease cognitive and motor functions and result in headaches and drowsiness. As with many air pollutants, effects may occur at lower concentrations for those suffering from other health problems. For example, a blood level of about 2 per cent carboxyhaemoglobin can affect some people with coronary problems, which is a blood concentration often found in smokers. The UK government has recommended an air quality standard of 10 ppm carbon monoxide, measured as an eight-hour running mean.

Nitrogen dioxide

Nitrogen dioxide can cause a number of respiratory problems, ranging from coughing through to aggravation of asthma. Asthma sufferers are the worst affected, with studies showing effects with exposures of about 100 ppb. As a result the UK government has recommended an air quality standard for protection of health of 104.6 ppb as a one-hour mean (with 20 ppb as an annual mean).

Experiments have shown that one effect of nitrogen dioxide is to reduce the ability of the cilia lining the bronchial passages to beat in a regular fashion. These hair-like structures carry mucus out of the respiratory system and so remove small particles and other contaminants, which may include allergens such as pollen. Thus exposure to nitrogen dioxide reduces this function.

Ozone

The health effects of ozone have been better studied in the United States than in other parts of the world, and much of this work has been focused on California where high ozone levels have prevailed for many years. Studies have included the experimental use of volunteers and animals and *epidemiological* studies of the general population and has covered both *acute*, short-term effects and longer-term impacts.

The studies have shown that a short-term exposure of at least 500 ppb for a few hours will cause inflammation of the respiratory tracts. Exposure to concentrations as low as 120 ppb for one hour can also increase sensitivity of some individuals to allergens such as pollen. In the UK, the Expert Panel on Air Quality Standards (EPAQS, 1994d) has concluded that exposure of people to 100 ppb ozone for several hours will cause inflammation of the respiratory system in sensitive individuals. Individuals are also likely to show a greater response when exercising, as the increased breathing increases exposure. With a safety factor incorporated, EPAQS has, therefore, recommended a standard of 50 ppb as an eight-hour mean.

Particulates

Not all particles have an equal impact on human health. As with many air pollutants, the main concern is with respiratory impacts. The *alveoli* in the deeper parts of the lung are characterised by thin, sensitive tissues (necessary to aid oxygen and carbon dioxide transfer) and are easily affected by pollutants. However, particles with a diameter greater than 10 μm are much less likely to penetrate deep into the lungs. Larger particles become trapped within the tracheal system and are removed in the mucus secretions.

Particles of less than 10 μm diameter are more familiarly termed PM_{10}. However, air monitoring records of PM_{10} are not strictly records of particles less than 10 μm diameter. A typical PM_{10} sampler will collect about 50 per cent of particles of 10 μm diameter, 95 per cent of those of 5 μm diameter and 5 per cent of those of 20 μm diameter. However, the exact concentration is less important than the need for consistency between measurements. If all monitoring and health studies are based on the same type of air quality data, then decisions, e.g. about standards, will be well founded.

Unlike many other air pollutants, no studies on controlled exposures on humans or animals have so far been undertaken with PM_{10}. All evidence leading to a concern over the presence of PM_{10} has, therefore, been derived from *epidemiological* studies. While these studies are advantageous in that they are based in the 'real world', it is often difficult to separate causative agents (EPAQS, 1995a). For example, high pollutant levels in urban environments (including particulates) often occur during the winter when meteorological conditions are unfavourable and pollutant emissions are highest. However, the cold weather itself can impact on health and separating pollutant effects from those of the weather can be difficult. EPAQS (1995a) also pointed out that monitoring of particulates in general and PM_{10} in particular has been very sparse, so that it can be difficult to relate good epidemiological data to adequate pollution information.

The World Health Organisation reviewed the evidence for impacts on health, and its conclusions are summarised in Table 2.8. However, interpretation of the information is not clear. For example, EPAQS (1995a) considered that as the increased mortality was found mostly in elderly patients with chronic pulmonary or coronary diseases, the effect of the PM_{10} was most likely to result on an earlier date for death, i.e. the actual effect may not be to increase mortality, but to change its timing.

TABLE 2.8 Estimates of the increase in occurrence of respiratory problems in asthmatic patients with changing concentrations of PM_{10}

Health indicator	Percentage change in indicator	Change in concentration ($\mu g/m^3$) in PM_{10} daily concentration required for the change in the indicator
Daily mortality	5	50
	10	100
	20	200
Hospital admissions for respiratory illness	5	25
	10	50
	20	100
Number of asthmatic patients using extra bronchodilators	5	7
	10	14
	20	29
Number of asthmatic patients noting increased symptoms	5	10
	10	20
	20	40

Sources: EPAQS (1995a) after WHO (1995)

EPAQS (1995a) concluded that the evidence was sufficient to indicate that there was a relationship between PM_{10} and health effects, and that increasing concentrations of PM_{10} resulted in increasing effects. However, it was unable to decide whether there was a threshold concentration below which effects did not occur. This was because, in part, compounding effects on health become increasingly difficult to separate out at lower concentrations.

As a result EPAQS (1995a) recommended an air quality standard for PM_{10} of 50 $\mu g/m^3$ as a daily average concentration. It acknowledged that this is a safe concentration for the majority of individuals, but there may be a few who will be adversely affected at lower concentrations. It concluded that efforts to reduce overall particulate levels, rather than just controlling peaks at or above 50 µg/m³ would bring the greatest benefits to the population as a whole.

Sulphur dioxide

Sulphur dioxide causes an irritant effect in the respiratory tract and results in coughing and tightening of the chest. The closure of airways may result and this is likely to be more common for those who suffer from asthma or those who may have lung diseases, for example those associated with smoking.

Experimental studies on healthy humans have shown narrowing of airways with exposure to sulphur dioxide at 4,000 ppb for five minutes. However, using asthmatic patients exercising to increase air intake, effects can be found at concentrations as low as 200 ppb or even less in particularly sensitive individuals. Sulphur dioxide may also increase sensitivity to allergens, but this is not yet proven. In the UK, therefore, EPAQs (1995b) has recommended a sulphur dioxide standard of 100 ppb over a fifteen-minute averaging period.

Heavy metals

Many metals are highly toxic. For example, while the public perception is that the threat of plutonium is as a strong radiation source (see Chapter 6), ingestion of even very small quantities is likely to be fatal – due to its toxicity, rather than from any radiological impact. Other metals, such as cadmium or nickel, *are* carcinogenic, so that their effects may take some time to become apparent.

Perhaps the best-known toxic-metal air pollutant is lead. The primary source of this pollutant is the use of leaded petrol, and public concerns over health impacts have led to some of the highest profile campaigns against any forms of pollution, and as a result the UK government has proposed an air quality standard of 0.5 µg/m³ as an annual mean. Lead is a cumulative poison, i.e. the main concerns are over long-term exposure to relatively low concentrations. Lead can affect the function of a number of organs, e.g. the kidneys. For many years lead effects were only recognised in their acute form, i.e. when *encephalopathy* was diagnosed. This requires a blood lead concentration of 80–100 µg/l. However, such concentrations only occur when significant quantities of lead are ingested or inhaled, e.g. from lead paint, and are not usually caused by general atmospheric pollution. However, more detailed studies have shown that at concentrations in the blood of 20–30 µg/l, effects on development of the central nervous system, kidney damage and impairment of haemoglobin synthesis will occur. Even lower concentrations will cause developmental problems in unborn children. Studies have found that lead poisoning can adversely affect children's behaviour and attention span and lead to lower IQs. The US Public Health Service estimates that blood concentrations of 100 µg/l are enough to cause long-term damage and a drop in IQ of two to four points. Surveys of children's blood in heavily polluted cities show levels much higher than this. For example, those in Cairo average 300 μg/l and those in Bangkok 400 μg/l. Using such data, the World Health Organisation has published guidelines for blood lead levels for adults and children (Tables 2.9 and 2.10).

TABLE 2.9 Lowest observable adverse effect levels for health effects of lead in adults

Blood lead level (μg/l)	Haem synthesis and haemotological effects	Effects on nervous system
1,000–2,000	—	Encephalopathic symptoms
800	Frank anaemia	—
500	Reduced haemoglobin production	Overt subencephalopathic neurological symptoms, cognitive impairment
400	Increased urinary Ala and elevated coproporphyrin	—
300		Peripheral nerve dysfunction
200–300	Erythrocyte protoporphyrin elevation in males	—
150–200	Erythrocyte protoporphyrin elevation in females	—

Source: WHO (1995)

TABLE 2.10 Lowest observable adverse effect levels for health effects of lead in children

Blood lead level (μg/l)	Haem synthesis and haemotological effects	Effects on nervous system
800–1,000	—	Encephalopathic symptoms
700	Frank anaemia	—
400	Increased urinary Ala and elevated coproporphyrin	—
250–300	Reduced haemoglobin synthesis	—
150–200	Erythrocyte protoporphyrin production	—
100–150	Vitamin D3 reduction	Cognitive impairment
100	ALAD-inhibition	Hearing impairment

Source: WHO (1995)

Impacts of air pollution on the natural environment

Acid rain

Of all the air pollution impacts on the natural environment, that of *acid rain* is the most publicised. Since the mid-1970s, its effects on forests and lakes have been regularly discussed by politicians and the media. While some claims of acid rain impact are now known to be erroneous, it is important to stress that exhaustive studies have well demonstrated extensive effects on many different parts of the natural environment, across many different countries. It is important to state this as there are some people in the media and in politics, at least in Britain and the United States, who claim that the widespread media coverage of acid rain in the mid-1980s was all generated by environmental lobby groups and has little scientific basis. Nothing could be further from the truth. The impacts of acid rain have been better studied than any other air pollution issue, including those of human health, and the evidence of impacts is now overwhelming. The twisted presentations of so-called scientific evidence by some anti-environmentalists must, therefore, be taken with a very large pinch of salt.

Acid rain can have some direct impacts on vegetation. For example, acid mists may damage the leaves of plantation forest trees. However, the most serious impacts occur via the general acidification of ecosystems. The most fundamental component of most ecosystems is the soil. Soils do have a natural buffering capacity against acidity. Their slow weathering releases base cations that neutralise the acid. However, the rate at which they can do this depends on the nature of the minerals present. Thus soils on limestone have an almost infinite capacity to deal with deposited acidity, while those derived from granite have very little capacity. In Britain the most sensitive soils occur in the uplands (Figure 2.1), which is unfortunate as that is where the highest levels of acid deposition also occur. In Europe, the largest area of sensitive soils occurs over much of Scandinavia, hence the concern with acid rain in Scandinavian countries. Similarly, the soils of the Canadian Shield are particularly sensitive and are downwind of much of the acid rain produced in the United States.

Acidification of soils has a number of effects. There is a loss of the base cations that buffer the acidity. These include plant nutrients such as calcium and magnesium, which may have important consequences for plant growth. In acid soils, aluminium takes over as the main buffering cation. However, aluminium is a toxin to both plants and animals. The ratio of calcium and magnesium to aluminium in soils has been found to be particularly important in determining the effects on the health of vegetation. The aluminium may also leach into surrounding waters. However, the different effects of freshwater acidification will be dealt with in Chapter 4. The link between soil chemistry and its ability to cope with deposited acidity is the basis for the policy analysis tool known as the *critical loads* approach, which is considered in Chapter 3.

Acid rain has two components – its acidity and, for nitric acid, the nutrient effects of the nitrogen. This latter effect will be included in the next section on ammonia (see p. 44). Acidity effects have been most described for effects on trees.

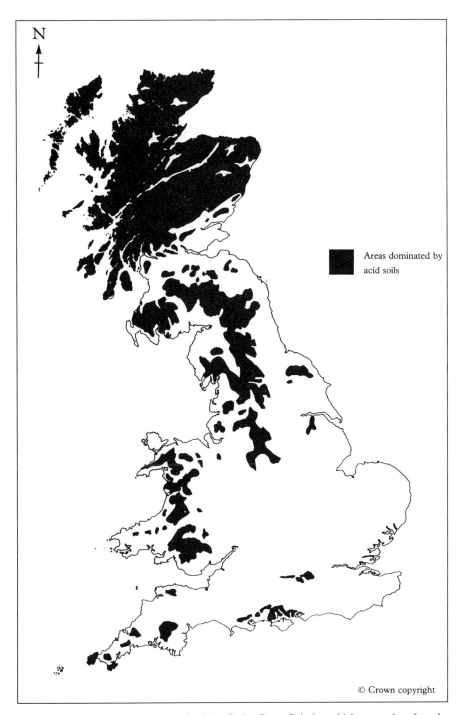

N

Areas dominated by acid soils

© Crown copyright

FIGURE 2.1 The distribution of acid soils in Great Britain, which may, therefore, be sensitive to acid deposition

Source: Skiba (1991). Reproduced with kind permission of the Joint Nature Conservation Committee.

Figure 2.2 outlines the mechanistic pathways that acid deposition might take in affecting tree health. Some of these are direct (e.g. foliar leaching) and others indirect (e.g. making trees more susceptible to drought or pathogens). It is important to note that the effects seen in one country may not be applicable elsewhere. For example, in Germany there was considerable concern over the effects of acid deposition on loss of magnesium (an essential element for chlorophyll production) from the soil, and its effects on tree health. However, in Britain the additional magnesium concentrations in rainfall from marine sources would mean that any leached magnesium should be

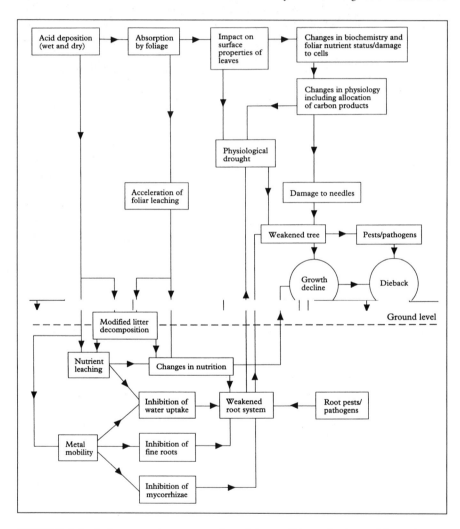

FIGURE 2.2 A diagrammatic representation of the possible mechanisms of forest damage that may be caused by acid deposition and thus lead to a decline in growth and dieback
Source: TERG (1988). Reproduced with permission of the Controller of Her Majesty's Stationery Office.

PLATE 2.2 A common way to examine how air pollutants affect plants is to grow them in Solardomes and expose them to different concentrations of pollutants. Alternatively one might take ambient air and filter it to see if current pollutant levels are having detrimental effects. Photograph: Andrew Farmer.

TABLE 2.11 Changes in defoliation of selected tree species between 1988 and 1991, resulting from surveys across twenty-eight European countries under the International Cooperative Programme on Assessment and Monitoring of Air Pollution Effects on Forests

	Percentage of trees in each defoliation class for different years											
	0–10 per cent				*11–25 per cent*				*>25 per cent*			
	88	*89*	*90*	*91*	*88*	*89*	*90*	*91*	*88*	*89*	*90*	*91*
Fagus sylvatica	59.8	60.8	54.0	50.5	26.0	27.5	28.5	31.3	14.2	11.7	17.5	18.2
Pinus sylvestris	60.8	62.9	58.9	54.1	26.9	27.2	27.7	29.0	12.3	9.9	13.4	16.9
Quercus robur	38.9	44.9	44.5	36.5	29.7	33.1	26.9	29.4	31.4	22.0	28.6	34.1

Source: CEC (1992)

readily replaced. There is no doubt that a range of factors affect tree health. However, trees are very difficult to study experimentally. Exposure to pollutant gases and acid deposition can only be undertaken with very young trees (for logistical reasons) and it is difficult to translate these results into impacts in the field. There are also good survey data, stretching over many years and for many countries, on the health of mature trees (Table 2.11). These show considerable variation in tree health from year to year and do reflect changes in the environment. However, generally it is difficult to disentangle the different environmental effects of drought, wind, pests and pollutants, especially as these often act in a *synergistic* fashion.

Acid rain affects a range of other habitats (Woodin and Farmer, 1993), including heathlands (Lee *et al.*, 1992), peatlands (Lee *et al.*, 1988) and epiphyte communities (Farmer *et al.*, 1992). The latter are interesting because in this case the acid effect is not mediated via the soil, but rather by acidification of the tree bark, which then affects the lichens and mosses growing on it.

Ammonia

In high concentrations, ammonia can cause direct damage to vegetation, resulting in burning and bronzing of leaves. Forests, for example, may show considerable damage in close proximity to intensive animal units (Roelofs *et al.*, 1985). However, at lower concentrations, ammonia is readily absorbed by leaves and does affect cell metabolism, due to the fact that the gas is alkaline and is a nutrient. When alkaline ammonia gas is deposited on to soil and into freshwater it leads to the apparently contradictory affect of acidification. This is due to the fact that the additional ammonia stimulates the activity of denitrifying bacteria, whose action results in the net release of hydrogen ions. The contribution of ammonia to acidification can be highly significant in some locations (e.g. heathlands in the Netherlands). However, it is the addition of nitrogen as a nutrient, which is the main pollutant problem arising from ammonia deposition. As this may have similar impacts from other forms of nitrogen deposition (e.g. the wet deposition of nitric acid and dry deposition of nitrogen oxides) general comments shall be made here.

Figure 2.3 outlines the various mechanistic pathways that nitrogen deposition may take when it affects vegetation. These mechanisms are focused at the level of the individual plant. The added nitrogen may decrease the need for mycorrhizal fungi and make the foliage more 'tender' and thus more palatable to insects and susceptible to frost. However, the biggest impact of nitrogen deposition is to change the competitive balance between species in a community. Many habitats are nutrient limited and the supply of additional nutrients will favour some species over others. Figure 2.4 shows the distribution of over 2,000 plant species from Central Europe, according to the nitrogen status of their habitats. It also shows the percentage of each group of species that are endangered. In every habitat type, the endangered species are always more likely to occur in low nitrogen conditions. Thus the addition of nitrogen from atmospheric pollution has the potential to threaten a wide range of rare species. Considerable work has been undertaken on the effects on a range of habitats. A review

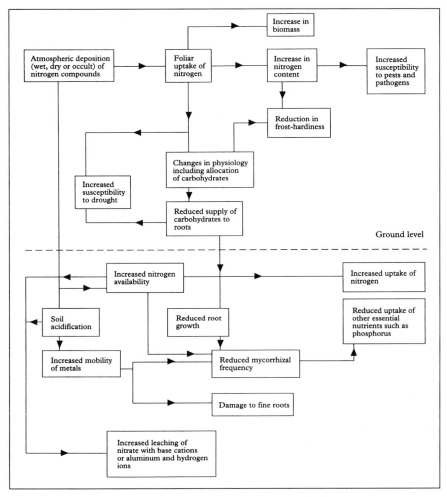

FIGURE 2.3   A mechanistic model of the soil and plant processes affected by nitrogen deposition

Source: Pitcairn (1991). Reproduced with kind permission of the Joint Nature Conservation Committee.

of these is beyond the scope of this volume, but the reader is encouraged to follow up these references for different habitats: chalk grassland (Bobbink, 1991), fungi (Jansen and van Dobben, 1987; Arnolds, 1991), heathlands (van der Eerden *et al.*, 1991; Lee *et al.*, 1992), uplands (Baddeley *et al.*, 1994) and woodlands (Tyler, 1987).

Fluorides

Hydrogen fluoride gas is toxic to many plants, with exposure causing various types of leaf injury and this will occur at very low concentrations. The same effects can be

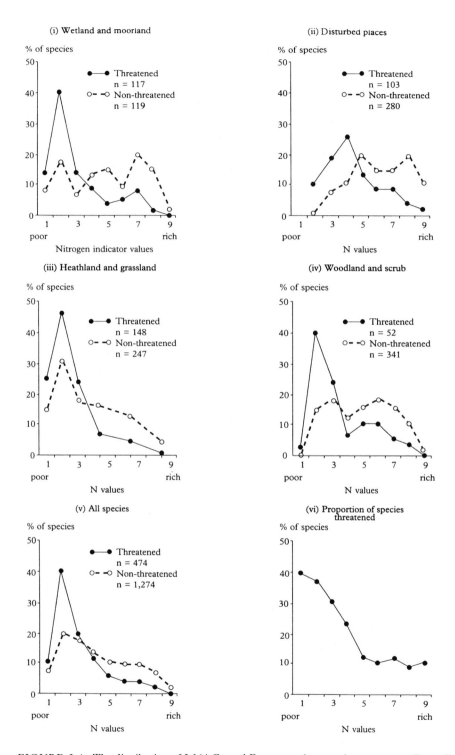

FIGURE 2.4 The distribution of 2,164 Central European plant species across a gradient of nitrogen indicator values, showing that, for a range of habitat types, threatened species predominate in nitrogen-poor situations

Source: Pitcairn (1991), after Ellenberg (1987). Reproduced with kind permission of the Joint Nature Conservation Committee.

produced by high soil fluoride concentrations, but these can be distinguished by examination of plant-tissue fluoride concentrations (high leaf fluoride concentrations indicate a direct atmospheric source).

Plants will accumulate fluoride from the soil. This can cause extensive damage to the plants themselves (e.g. lowering of crop yields), but of greater concern are the effects that this may have on the animals that eat those plants. Domestic and wild animals suffer from fluorosis, which has a number of symptoms, including tissue swelling, lethargy, bone disease, wasting and, finally, death. In particular, fluoride competitively interferes with the metabolic processes involving metals within the body, such as calcium and magnesium, and the growth regulation associated with these metals, such as parathyroid activity which regulates bone tissue production. The fluoride absorbed can also be passed on to humans. It is rare to find extreme toxicity effects in humans. However, there is concern over long-term chronic exposure, hence the debate over the fluoridation of drinking water for dental protection.

Nitrogen dioxide

Nitrogen oxides are important contributors to acid rain and nitrogen deposition. However, exposure of plants to gaseous nitrogen oxides can cause direct toxic effects (Steubing *et al.*, 1989; Wellburn, 1990). The short-term exposure of plants to nitrogen oxides needs relatively high concentrations to affect photosynthesis, for example Bennett *et al.* (1990) found that 380 ppb nitrogen dioxide were required before photo-synthetic inhibition occurred. However, this effect could be found at only 80 ppb when sulphur dioxide was also present. This *synergistic* effect of combinations of air pollutants is important, but too often rarely studied.

However, long-term exposures to much lower concentrations do have adverse effects on the growth and yield of crops (e.g. Sandhu and Gupta, 1989) and wild species (e.g. Ashenden *et al.*, 1992). The former work found low concentrations resulted in stimulation of growth, presumably by a fertilising effect. While this might be beneficial for crops, it is considered unacceptable for native species due to the potential changes in the competitive balance of plant communities.

Ozone

Ozone can have a wide variety of effects on the natural environment. At relatively low concentrations, crop yield can be reduced even when no visible injury can be seen. Crops found to be affected include barley, beans, clover, peas, spinach and wheat. Recent studies have also found extensive impacts on crops on the Indian subcontinent, where ozone levels are also high. Apart from the direct damaging effects of ozone, it has also been found that the growth rates of some aphids may be higher on plants exposed to ozone and this may also affect yield (Brown *et al.*, 1992).

Certainly ozone can affect the health of young trees, under experimental conditions (PORG, 1993), and growth studies have shown that some tree species will grow better in ambient air when ozone is filtered out (Durant *et al.*, 1992). It is, however, difficult to assess whether ozone is affecting the health of forests. While, for example, the health of some trees is poor, this is often linked to drought or pests. However, drought conditions (dry and sunny) are the same as those which produce ozone and some effects of ozone and drought on tree seedling health are *synergistic*. Separating these effects is therefore very difficult.

Ozone has been shown experimentally to affect a wide range of native herbaceous species (Reiling, 1992). However, care has to be taken when interpretating short-term single-species experiments in the field. From long-term studies of a grassland community Evans and Ashmore (1992) found that ambient ozone levels do have significant effects, but these were not predictable from previous, single-species, studies.

Particulates

Particulates or dusts can have considerable local impacts on the natural environment. However, it is important to note that the different types of dust are not equivalent in their impact and that two characteristics have to be considered. These are size and chemistry.

Particles from motor vehicles may range from 0.01 to 5,000 μm diameter, while those from cement kilns are generally less than 30 μm (Farmer, 1993b). Larger particles are deposited more rapidly. Thus Everett (1980) found that particles of more than 50 μm were all deposited within 8 m of an unpaved road. Dust size not only affects dispersion, it also affects impacts. Many dust effects on vegetation occur via blocking of leaf stomata: the pores through which plants undertake the gas exchange necessary for photosynthesis and respiration. The size of the stomata does vary, but is generally in the range of 8–12 μm. Dusts of this size range are therefore likely to have greater impacts.

In sufficient quantities, dusts may form a smothering layer on leaves, reducing light and hence lowering photosynthetic rates. Many dusts are inert and so only act by shading. However, some dusts are also chemically active. Thus cement dust will also dissolve leaf tissue, resulting in additional injury. Coal dust may also contain toxic compounds. Dusts may also affect ecosystems through their action on soil. Thus the alkaline chemistry of limestone dusts can raise the soil pH of acid and neutral habitats, resulting in the loss of plant and animal species.

A wide range of dust impacts have been recorded, although there are many gaps in our knowledge. These impacts have been reviewed by Farmer (1993b) and are summarised in Table 2.12.

The degree of impact is also affected by the ease with which dust will deposit on to plants. Those species with diffuse structures tend to trap dust more efficiently. Thus mosses may gather large quantities of dust, which is difficult to wash off during rain. Many coniferous trees have dense needle foliage that may trap dust efficiently. While this may give an indication of the species more likely to be at risk, it also provides some guidance to the pollution manager in attempting to manage dust

TABLE 2.12 Examples of the types of impacts that have been recorded on vegetation and communities for different dusts

Dust type	Effects on the natural environment
Cement	Blocked stomata
	Reduced fruit set, leaf growth, pollen growth
	Leaf necrosis and chlorosis, bark peeling
Limestone	Blocked stomata
	Reduced tree growth, except for occasional tolerant species
	Changes in soil chemistry
	Loss of species such as heather, sensitive mosses and lichens
	Effects on invertebrate populations
Road	Blocked stomata
	Increased leaf temperature
	Reduced photosynthesis
	Change in habitat species composition, loss of mosses and lichens
Fertiliser factory	Increase in young tree growth
	Leaf injury

dispersion. Some coniferous hedge species, for example, are very fast-growing and can act as good dust traps, and, being evergreen, this effect lasts all year. These species could, therefore, be used as screens near quarries for example.

Sulphur dioxide

There have probably been more studies on the direct effects of sulphur dioxide on plants than any other pollutant. However, many of these studies are now rather old and reflect the short-term analyses of the impacts of very high concentrations. Today, sulphur dioxide levels have been considerably reduced in almost all parts of the world, apart from some areas of the developing world. Thus many published studies do not help in understanding current potential impacts.

The group of organisms most sensitive to sulphur dioxide are lichens. These species readily absorb the pollutant and it is highly toxic to their metabolic function. Most species are so sensitive that historically high levels of sulphur dioxide have caused the loss of many species throughout large parts of Europe, and many urban areas have become known as 'lichen deserts'. The response of different species varies, and this has enabled researchers to produce lichen maps of countries and regions which correlate with their sulphur dioxide levels (e.g. Hawksworth and Rose, 1970). These maps have been used as biomonitors of air pollution for some time (e.g. Figure 2.5). However, as pollutant levels have dropped, some species have re-invaded, although not always in the order in which they disappeared (Gilbert, 1992). Also

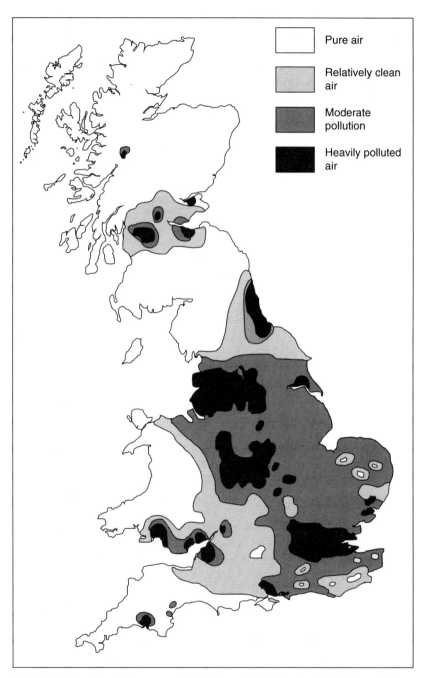

Pure air

Relatively clean air

Moderate pollution

Heavily polluted air

FIGURE 2.5 Air pollution zones in Britain, as defined by the distribution of lichens
Source: Richardson (1975). Reproduced from *The Vanishing Lichens* (David & Charles, 1975)
with kind permission of the publishers.

many mature trees have been so badly affected by air pollution that their bark is no longer suitable for lichen colonisation (Bates *et al.*, 1990). There is, therefore, a limit to the usefulness of lichen mapping as a biomonitor in declining sulphur dioxide environments (although it remains useful as an analysis of how this component of our biodiversity is doing). It is also curious to note that while most lichens are very sensitive to sulphur dioxide, there is one species, *Lecanora conizaeoides*, which is dependent on the presence of sulphur dioxide and which has spread throughout Europe over the last 150 years. It is thought to be native to the volcanic, sulphurous areas of Iceland and has shown a recent decline as air pollution control measures have been adopted.

The effects of sulphur dioxide can also been seen on many higher plants, including wild species, crops and trees. Effects include cell metabolism disruption (membrane damage, respiration and photosynthetic effects), leaf injury and loss, reduced growth and reproduction. In many cases it is not always clear how significant these effects are. For example, a transitory effect on a photosynthetic measurement is a response, but in the field may have no real impact on the long-term survival of the plant. However, effects on growth are probably highly significant. The effects of sulphur dioxide on plants are considered in further detail by Bell (1994), Dueck *et al.* (1992) , Steubing *et al.* (1989) and van der Eerden and Duijm (1988). However, it is important to note that some species do develop sulphur-dioxide-tolerant populations in response to long-term exposure (Ayazloo and Bell, 1981), although the effects on genetic diversity are not known. Sulphur dioxide may also increase the susceptibility of plants to attack by insect herbivores (see for example Oksanen *et al.*, 1996).

Impacts of air pollution on the built environment

The acid gas pollutants (sulphur dioxide, nitrogen oxides and their oxidation products) and ozone all have impacts on buildings and materials. In many instances the impact on our cultural heritage is one of the most visible effects of air pollution. Throughout most of Europe it is possible to walk through many cities and see the erosion to historic artefacts – crumbling stone, and statues on cathedrals that are no longer recognisable. Even archaeological monuments, such as ancient rock carvings in southern Sweden, have been damaged. Indeed effects on buildings can be seen world-wide. In 1993, concern over the yellowing of the white stone of the seventeenth century Taj Mahal in India led the government to close 200 nearby factories and a coal-fired power station. An oil refinery 80 km away is also under threat of closure.

Sulphur dioxide deposits on to stone surfaces and is oxidised to sulphate, producing hydrogen ions. This acid production may then be neutralised by the limestone in buildings, resulting in the production of calcium sulphate, which is readily lost as erosion occurs. Nitrogen dioxide acts in a similar manner, but its deposition rate is much lower than for sulphur dioxide. Ozone acts on different materials, such as rubber, textiles and paint. Its strong oxidising action can rapidly degrade these substances. In the UK it is estimated that ozone impacts alone result in annual damage costs to buildings and materials of £350 million. Experiments have also found that these three pollutants have a strong *synergistic* effect, with combinations of two

TABLE 2.13   Damage caused to different types of material by different air pollutants

Material	Type of damage	Main air pollutants
Building stone	Surface erosion Discoloration Crusting	Sulphur dioxide Nitrogen oxides Deposited acidity
Metals	Corrosion Tarnishing Pitting	Sulphur dioxide Hydrogen sulphide Other acid gases
Paints	Discolouration Peeling Cracking	Sulphur dioxide Hydrogen sulphide Alkaline aerosols Deposited acidity Photochemical oxidants
Ceramics and glass	Erosion Crust formation	Acid gasses Fluorides
Rubber	Cracking	Photochemcial oxidants

Source: Yocum and Baer (1984)

of the pollutants having a greater effect than their mere addition would suggest. Table 2.13 shows the types of damage, and the main pollutants causing this damage, for the different types of material.

The impacts of air pollution on materials are heavily influenced by meteorology. Buildings in different parts of a town may exhibit very different degrees of damage. Those buildings exposed to rain will tend to show erosion effects, while those in sheltered areas will develop crusts on their surfaces. Buildings close to particulate sources (e.g. roads) are more likely to be discoloured.

It is important to assess the degree of damage that different concentrations of air pollutants cause to buildings and materials, in order to devise effective control strategies. The United Nations Economic Commission for Europe, for example, coordinates a series of sites throughout Europe, where materials of different types are exposed to the atmosphere, and air pollutants are monitored. Correlations between degree of damage and concentrations of pollutants can be compared between sites, and critical loads for pollutants devised. In the UK the exposure network is expanded to form the National Materials Exposure Programme (NMEP) (Butlin et al., 1995).

The NMEP programme seeks to derive dose–response relationships for the decay of different materials in a general form:

$$\text{Decay rate} = \text{constant} + \text{constant} \times [SO_2] + \text{constant} \times [NO_2] + \text{constant} \times [\text{acidity}] + \text{constant} \times [\text{rainfall}]$$

Many materials (e.g. limestone) do slowly decay in clean air, for example due to the presence of carbonic acid in rain. However, by establishing a decay rate which is deemed acceptable, it is possible to use the above relationships to identify pollutant concentrations that produce acceptable impacts.

Impacts on visibility

Visibility is not easily defined, yet it is well appreciated by the general public. Haziness is caused by light transmission within the atmosphere being attenuated, either by particles or by certain gases. The effects of air pollution on visibility are also well publicised. The most notorious of these were the London smogs (the 'pea-soupers') from the late nineteenth century to the late 1950s. These were caused by the build-up of high concentrations of particulates produced from domestic coal combustion. Currently the most common impact on visibility is produced by photo-chemical smogs. Here oxidants such as ozone can cause conditions where light scattering occurs. As such oxidants are very common in rural areas, an impact on scenic views often occurs. The particles and gases affect visibility in three ways:

1   The light travelling from an image is scattered so that it never reaches the observer's eye.
2   The light travelling from an image is absorbed and does not reach the observer.
3   Background light is scattered into the sight path of the observer, blurring the image.

However, for all that has been experienced in Europe in terms of visibility impact, it is in the US that most concern has been expressed and most direct effort has been made to solve the problem. The focus here has been on national and state parks where visibility is explicitly given as a major reason for visiting the park. Thus the US National Park Service (NPS) has established a visibility monitoring programme, which shows that for more than 90 per cent of the time scenic views are affected by anthropogenic pollution at all the monitoring locations in the forty-eight mainland states (NPS, 1988).

The US NPS has adopted a measurement scale to assess visibility impacts. Individuals are asked to judge visibility on a one to ten scale termed 'perceived visual air quality'. Rather than a simple measure of pollutant levels or light transmissibility (the monitoring of which is also undertaken), this scale incorporates the subjective requirements of visibility, which is what is actually requiring protection. The results from these studies have been found to show reasonable consistency.

Studies have shown that in areas of the US where there is little air pollution (e.g. rural and desert areas of the south-west) the visibility averages between 130 and 190 km. However, in the more-polluted east this drops to around 20–35 km. In both the east and west, sulphate aerosols are the most common causes of visibility problems. Comparison of visibility data with pollution data across different states and over time has led to the development of models that can assess the benefits of different pollution reductions. Thus a reduction in sulphur dioxide emissions of 10 per cent (and hence a reduction of 10 per cent in the sulphate aerosols that scatter the light) will improve visibility by 6 per cent (NAPAP, 1992).

The US has also adopted specific air pollution controls aimed at improving visibility. As far back as the 1977 Clean Air Act, the US defined 'Federal Class 1' areas (e.g. National Parks) which required protection. As a result, the US Environmental Protection Agency has targeted specific pollutant sources close to these areas. The aim of such regulation was clarified in the 1990 Clean Air Act, which stated that there should be 'reasonable progress towards the national goal of no man-made visibility impairment' in the Class 1 areas.

It is interesting that some of the earliest studies of the public's willingness to pay for air pollution reductions were undertaken on visibility effects in the US. In 1980 the NPS undertook a study in the Grand Canyon and Mesa Verde National Parks. It found that visitors would be prepared to travel further to enjoy clearer views. An average visitor party would have been prepared to pay $1 to have a view of 160 km rather than 110 km, and $3 for a view of 350 km. While this might seem a small sum, it should be noted that in 1980 the number of visitors to the Grand Canyon was 2,305,000, with an average of three per party, and therefore represents a significant financial interest.

Economic impacts

The effects on buildings and health are important in assessing overall strategies to reduce air pollution. This is because these impacts have direct economic costs. The economic impacts of effects on the natural environment are much harder to quantify. A number of studies have examined the economic impacts of current air pollution levels and the effects of control options. For example, Mayerhofer *et al.* (1995) estimated that current levels of sulphur dioxide and nitrogen oxide pollution in Europe causes between 2.9 and 5.3 billion Ecu per year in damage to buildings alone. By assessing this in terms of individual fossil-fuel power-station function, this was equivalent to between 0.0062 and 0.12 Ecu per megawatt hour.

The UK Department of the Environment commissioned work to examine the costs and benefits of sulphur dioxide abatement (ENDS, 1995a). The study examined the economic benefits of sulphur dioxide abatement plans for the UK up to 2030, assuming compliance with agreements under the 1994 UNECE Sulphur Protocol. It assumed a standard 6 per cent discount rate and used economic analyses of market values and, where necessary, *contingent valuation*. Table 2.14 presents the results for different sectors of the environment affected by sulphur dioxide pollution. The study

TABLE 2.14 The present value of economic benefits resulting from the UNECE sulphur dioxide control strategy for the UK

Environmental receptor	Lower bound ($£m$)	Central estimate ($£m$)	Upper bound ($£m$)
Vegetation	0	29	164
Aquatic ecosystems	0	711	3,996
Crops	11	19	19
Forests	0	0	1
Human health	505	14,595	14,595
Modern buildings	1,914	2,614	3,566
Total	2,430	17,968	22,341

provided different estimates for each example, resulting in a central estimate with lower and upper bands.

The main economic benefits are, therefore, to be found in human health and buildings. However, the study cannot take account of non-monetary values, e.g. of ecosystems. The central estimate of £18 billion demonstrates the large economic impact that air pollution can have, and while estimates of costs of meeting sulphur pollution control measures are large (£1–3 billion), these are considerably less than the costs of non-compliance.

In order to link impact costs with those for controls it is better to present these impacts in relation to the quantity of pollutants that are emitted. Thus in the UK it is estimated that one tonne of sulphur dioxide will cause damage costs of £1,329–2,372. One tonne of particulates will result in overall costs of about £3,680 and that for nitrogen dioxides of £720–805.

Chapter summary

♦   There is a wide range of air pollutants. This chapter examines the sources of a number of these and the impacts they may have on human health and the natural and built environment. These pollutants include: ammonia, carbon monoxide, fluorides, heavy metals, nitrogen oxides, ozone, particulates, sulphur dioxide, organic toxins, volatile organic compounds and wet deposited acidity ('acid rain').

♦   Health effects may be divided into acute and chronic impacts. Thus a smog may cause immediate acute suffering to those with asthma, while long-term exposure to lead may cause the chronic effect of mental impairment in children. A range of pollutants have impacts on respiratory function, with related coronary effects, e.g. nitrogen oxides, ozone, particulates and sulphur dioxide. Others, e.g. carbon monoxide or heavy metals, are toxins. Most recently, concern has been expressed over small particulates ($PM_{10}s$). In managing the effects of pollutants on human health, there is great difficulty in defining the pollutant levels that are 'safe'. Many standards are not set at a concentration considered to have 'no effect', but where effects are 'negligible'. There is therefore a strong interaction between scientific analysis and social values.

♦   There is also a wide range of impacts on the natural environment. The most publicised and well-documented are those of acid rain, causing acidification of soils and freshwaters, with direct impacts on a number of species, such as fish, and indirect effects on species whose food may disappear, e.g. birds such as the dipper. An increasing problem is that of nitrogen deposition (from ammonia and nitrogen oxides), leading to the eutrophication of ecosystems and threatening many rare species. Some pollutants (e.g. fluorides or dusts) may have local effects around specific sources; these demand different management responses.

◆ Many buildings are at risk from air pollution. Acidifying pollutants (e.g. nitrogen oxides, sulphur dioxide and acid rain) may dissolve limestone and erode building-stone and historic artefacts. Particulates may discolour buildings, and ozone attacks other materials such as rubber or paint.

◆ Air pollution also affects visibility, which is especially valued in the country-side. This has been studied extensively in the US and this chapter describes the changes in visibility that have occurred in National Parks there, and the controls set in place to reduce the problem.

◆ Air pollution, through all of its impacts, may have direct or indirect economic effects, e.g. the need to repair buildings or days taken off sick by those with health problems. An example is given of a study to examine the economic benefits resulting from controls to reduce sulphur dioxide emissions in the UK.

Recommended reading

Department of the Environment Review Group Reports. The UK DoE operates a series of independent review groups examining different air pollution problems. Each examines current research to summarise the science and assess policy implications. Each report is an excellent introduction to individual topics. Many of the reports are referenced at the end of this volume. Current review groups include:

Acid rain review group
Acid waters review group
Building effects review group
Critical loads advisory group
Expert panel on air quality standards
Impacts of nitrogen deposition in terrestrial ecosystems review group
Photochemical oxidant review group
Terrestrial effects review group
Urban air quality review group.

Elmsom, D. (1996). *Smog Alert: Managing Urban Air Quality*. Earthscan, London. This is an excellent review of the sources of air quality problems in urban areas, and the effects that these may have on human health. It details a wide range of management options, taking account of experiences in Europe, North America and the developing world.

Wellburn, A. (1995). *Air Pollution and Acid Rain: the Biological Impact*. 2nd edn. Longman, London. This is an excellent, detailed account of the wide range of air pollution impacts on the natural environment, covering both the effects of ambient air quality and of deposited pollutants.

Air pollution, regulation
and management

◆ Introduction   58
◆ Air pollution regulation in the United
   Kingdom   58
◆ Air quality standards in the United Kingdom   63
◆ Air pollution regulation by the European
   Union   70
◆ The United Nations Economic Commission
   for Europe (UNECE)   75
◆ The critical loads approach: assessing national
   and international policies   77
◆ Monitoring air pollution   79
◆ Modelling air pollution dispersion   82
◆ Techniques to reduce air pollution   87
◆ Environmental assessment   99
◆ Chapter summary   100
◆ Recommended reading   101

Introduction

Given the wide range of air pollutants and their sources and the large number of impacts that these have, it is important that comprehensive and sophisticated systems of regulation and management are adopted. Over time, these systems have indeed developed, and this chapter will describe them. Initially it provides an examination of regulation in the UK. This includes the system of Integrated Pollution Control, which also controls emissions to freshwaters and the seas. The international context (EU and UN) is also extremely important, as air pollution is often trans-boundary. Management of air pollution requires a range of monitoring and modelling techniques, and these are described. Finally, the wide range of management options for control of air pollutants in three areas: industry, agriculture and traffic, are discussed.

Air pollution regulation in the United Kingdom

Air pollution regulation has a very long history in the UK. Combustion processes emitting air pollutants have been used for thousands of years. For example, coal was used by Iron Age man for smelting iron ore. The consequences of such pollution are also possible to detect. Coal burning in medieval York resulted in a number of classic diseases that are retained in skeletal remains (Anon., 1995). The first recorded regulation actually dates to 1273, when a prohibition on the use of coal was passed in London. This legislation was not effective and the problem continued to grow. One response in the seventeenth century mirrors the 'tall-stack' policy of the 1950s and 1960s, with a baker being ordered to use a chimney of sufficient height to allow the smoke to clear his neighbours' houses.

The first 'modern' legislation was the series of Alkali Acts of 1863, 1874 and 1906 (with some minor intermediate Acts). These Acts introduced a range of principles still present in current legislation. Controls were placed on offensive emissions. They were also not limited to chimneys, but also referred to *fugitive emissions*. Processes were linked together in groups for ease of administration. The 1874 Act introduced the first statutory emission limit (for hydrogen chloride) and also the concept of 'best practicable means' (BPM) for controlling emissions. The Health and Safety at Work Act 1974 set up a system of registration, the use of BPM and inspection. Processes could be issued with enforcement notices if the legislative provisions were not being complied with.

Controls over the nuisance created by emissions in urban areas came to a head in 1952, when a severe London smog lasted for five days in December, resulting in

PLATE 3.1 A typical British coal-fired power station. Note that all of the pollutants that are of concern are discharged from the two tall stacks to the right. It is a common public misconception that the large cooling stacks to the left, with huge billowing clouds of steam, are emitting pollutants.
Photograph: Andrew Farmer.

4,000 additional deaths. The public outrage resulted in the passing of the 1956 Clean Air Act (amended in the 1968 Clean Air Act). This controlled smoke emissions, introduced 'smokeless zones' in many urban areas, and required increased stack heights for many industrial processes. The emission of sulphur dioxide higher in the atmosphere did solve many of the local smog problems, but only added to the problem of long-distance transport and the deposition of air pollutants as acid rain.

Integrated pollution control in the United Kingdom: the 1990 Environmental Protection Act

The UK's current system of pollution regulation is based on the 1990 Environmental Protection Act, and is administered by bodies created in the 1995 Environment Act. The Environmental Protection Act set up a system of integrated pollution control (IPC). Prior to this, the regulation of pollution to land, air and water was dealt with under separate systems. No consideration was given as to whether discharges to one medium could lead to less environmental harm if discharged to another. For example, is it better to landfill certain types of waste (with implications for impacts of land-take, leachate and landfill gas) or to incinerate it (with potential air pollution impacts)?

In 1976 the Royal Commission on Environmental Pollution recommended a system that would lead regulators to decide on the 'best practicable environmental option' (BPEO), i.e. to manage waste and pollution to result in the best overall benefit for the environment. This was adopted in the Environmental Protection Act under the system of IPC, and is enforced through three bodies – the Environment Agency (England and Wales, combining the former Her Majesty's Inspectorate of Pollution or HMIP, the National Rivers Authority and the Waste Regulation Authorities), Scottish Environmental Protection Agency (SEPA) and The Environment Service (Northern Ireland). IPC has the following features:

♦   It covers all of the most seriously polluting processes which emit air pollutants, discharge specified dangerous substances to water or generate significant quantities of 'special wastes'.

♦   Each operator must have authorisation prior to discharging pollutants.

♦   In granting authorisation the authority must enforce existing emission standards and not allow a breach of any statutory environmental quality standards.

♦   The authority must then require the application to use 'best available techniques not entailing excessive cost' (BATNEEC) to prevent or minimise release of substances and to ensure that the release of substances does not harm the environment. Harm is defined in the Act as 'harm to the health of living organisms or other interference with the ecological systems of which they form a part and, in the case of man, includes offence caused to any of his senses or harm to his property'. This is a very comprehensive statement of environmental protection.

♦   The authorisation process involves formal input from other interested agencies (e.g. the Health and Safety Executive, nature conservation bodies, etc.) and, via advertising, input from the public.

The aim of preventing 'harm' has been taken seriously by regulatory authorities in the UK. For example, HMIP as criticised for delays in defining future sulphur dioxide emission limits for the coal- and oil-fired power stations in England and Wales. However, when the authorisations were finally given in March 1996, they contained a requirement to reduce emissions by about 85 per cent from 1996 levels, by the year 2005, stipulating that most of the reductions should occur by 2001. This is significantly more than would be required by the UK's international commitments for sulphur emissions. However, the limits were set by assessing whether further harm could be prevented by increasingly strict controls. The limits set reflect a point where further reductions in impacts would realistically only be achieved by controlling other sulphur sources, such as those in the rest of Europe.

Not all pollutant discharges are regulated under IPC. Smaller processes emitting air pollutants are regulated under Local Authority Air Pollution Control (LAAPC). This is operated in England and Wales by District Councils (or unitary authorities), and in Scotland by the SEPA. For these processes, emission, monitoring and housekeeping standards are set nationally, and the authority is only required to ensure compliance with these standards. Some important air pollution sources (e.g. ammonia from agriculture) are not regulated, but are covered by non-statutory codes of practice, although some of these may be brought under regulation by a new EU Directive (see p. 72). Many important aquatic discharges are also not covered by the IPC, e.g. discharge from agriculture and sewage treatment works. These are subject to separate regulatory systems. Similarly, only a fraction of waste discharges (by weight) are covered by IPC.

The Environment Agency also has an important role in informing the public of pollutant releases. All significant releases from regulated processes are recorded in a Chemical Release Inventory. This is accessible over the Internet by the public who can interrogate the CRI in different ways, e.g. total releases in a county or by industrial sector.

CASE STUDY: PROSECUTION OF COALITE CHEMICALS' INCINERATOR, BOLSOVER, DERBYSHIRE

The case of Coalite Chemicals' plant at Bolsover, Derbyshire, England is an interesting example of the interaction of many players in the management of environmental pollution, involving regulators, industry, landowners, environmental groups and the general public. The case was concluded in February 1996 when the company pleaded guilty in a prosecution brought by HMIP and a record fine of £150,000 and £300,000 in costs was awarded (ENDS, 1996). However, the case had a long history.

The plant under consideration was a chemical waste incinerator, burning large quantities of waste. Concern had been expressed about the plant operation for many years, but two significant findings proved of particular concern. In mid-1991, the Ministry of Agriculture, Fisheries and Food found that grass, cattle and milk from farms in the immediate vicinity of the incinerator were contaminated with dioxins (ENDS, 1993a). As a result, on two farms, milk sales were suspended, and on a third, farm sales of contaminated suckling cows were suspended. There was widespread public anxiety and a series of parliamentary questions resulted. The company was the subject of much public acrimony. The National Farmer's Union, acting on behalf of the farmers affected, prepared to take Coalite to court to seek compensation. However, before the case could reach court, an out-of-court settlement of £200,000 compensation was agreed in late 1993. In settling the case, the company did not, however, accept liability (ENDS, 1993b). The NFU were also critical of HMIP for not releasing details of the sampling they had undertaken around the plant. However, these data were being withheld as HMIP considered their own prosecution.

Following the earlier finding of contamination of farmland, the NRA examined the river Doe Lea, into which the Coalite works had a discharge consent (ENDS, 1995b). Samples of sediment taken in 1991 just downstream of the plant, showed the highest dioxin levels ever reported in river sediments, i.e. 64,000 ng/kg of dioxins expressed as a toxic equivalent (TEQ) of the most toxic dioxin: 2,3,7,8–TCDD.

Average levels for a river receiving industrial discharge are about 15–50 ng/kg TEQ. However, the contamination was not limited to the receiving river, and sediment samples from the River Rother, 20 km downstream, showed concentrations of 1,700 ng/kg – still about 100 times above background levels and ten times higher than any other river in the UK.

In November 1991, the incinerator was closed down voluntarily. Following this, the levels of dioxins on the surrounding farms fell, with a decline in their concentrations in milk and in the farmers' blood (in those who drank the milk) and also levels in the river.

The NRA considered a criminal prosecution. However, it was thought that this would be unlikely to be successful, as the existing discharge consent was old and poorly worded. There were few conditions on the discharges, and no specific limits on dioxins. A revised consent, which would have covered dioxin emissions did not come into force until 1993 – after the incinerator was shut. The NRA then considered a civil prosecution to obtain costs for cleaning the contaminated river Doe Lea. It estimated that the clean-up would cost around £1 million. However, further examination of the contamination suggested that removal of the sediments might result in more of a threat to humans and the wider environment, due to their disturbance, than would be the case if leaving the sediments intact. Without any incurred costs, the civil case could not proceed.

This left the only option for prosecution to HMIP. Following the concerns over farm contamination, HMIP undertook some stack sampling on a single day in August 1991, and analysed the chemical waste prior to burning. It was found that stack emissions of dioxins and furans were 88 and 36 ng/m^3 and that waste concentrations ranged from 200–1,000 ng/kg. In early 1993 Greenpeace entered the works secretly and took waste samples (Greenpeace, 1993). Analysis of these found dioxin levels about nine times higher than those collected by HMIP. Greenpeace severely criticised HMIP as a result.

HMIP, however, continued its own investigations. During early 1993, lengthy site visits were made by the inspectors, to examine log books and interview staff (Nicholson, 1996). They concluded that the incinerator had not been operated in accordance with earlier requirements, and thus not in accordance with best practicable means. A prosecution was deemed likely, but further sample examination was required. In early 1994, samples from the plant were re-analysed, and it was concluded that the earlier analysis was in error, and high dioxin levels were found in dusts from ductwork, and in surrounding soil samples. The analysis also examined the dioxin and furan 'fingerprint'. These chemicals belong to a group of 210 substances, and any emission source will have a particular combination of chemicals characteristic of that source. The same 'fingerprint' was found in the feed waste, gas emission samples, dust, grass and soil samples. With a chemical link between the emissions and environment, and the decline in the environment following incinerator closure, HMIP felt that it had a strong case.

The case revolved around written instructions given by HMIP in 1989, whereby the outlet chamber from the incinerator was to be kept above 800°C to ensure destruction of the dioxins. Even this constraint might be deemed somewhat lax, as the same year, HMIP issued a guidance note requiring incinerators burning chlorinated waste to operate at temperatures of 1,100°C. However, examination of the Coalite incinerator activity showed that it did not even comply with the 800°C requirement. Between 1989 and 1991 nine incidences of non-compliance were found. One of these lasted for four days. An incident even occurred in August 1991, after problems on local farms had become apparent.

Coalite was therefore charged with a breach of section 5 of the Health and Safety at Work Act 1974, for failure to use best practicable means to prevent the emissions. In early 1996, Coalite decided not to contest the case and pleaded guilty. The failure to ensure the maintenance of the incinerator temperature received particular attention from the judge, who said:

> In my judgement it is not only the operatives and their supervisors who are to blame for what happened. I accept that the company caused written instructions to be posted in the control room of the incinerator in April 1990, and that these instructions were disobeyed on the occasions I have mentioned, but I remain of the view that it is the management of the company who are most seriously to blame for not ensuring that their staff knew why they were being instructed not to burn waste within 12 hours of starting up, or at temperatures of less than 800 degrees, and for not having effective monitoring systems in place to ensure that this did not happen . . . they were obliged to be vigilant to see that the law was obeyed and the local environment protected.

The case demonstrates the painstaking work that needs to be undertaken in bringing a prosecution. It shows how older consents to pollute can provide loopholes preventing prosecution. The valuable role of public pressure and that from environmental groups is highlighted, and the ultimate responsibility of company managers for what their employees do is stressed.

Air quality standards in the United Kingdom

A key component of air pollution regulation in Europe, the UK, and a number of other countries, is that of the use of air quality standards. Their role in the UK was expanded by the government in its National Air Quality Strategy (DoE, 1997). The aim of the strategy is to tackle pollution 'hot-spots' and institute air quality management procedures to deal with these. Such hot spots are to be identified by targeted monitoring, with the results to be compared with a new suite of air quality standards.

The standards for each of the nine pollutants (Table 3.1) have been set following recommendations from the government's Expert Panel on Air Quality Standards (EPAQS). The EPAQS was set up to assess the scientific and medical information for the effects for each pollutant, and to recommend concentrations that would 'minimise risks of harm'. Within the National Air Quality Strategy, the UK

TABLE 3.1 Standards recommended for the UK by the Expert Panel on Air Quality Standards together with the proposed government targets to achieve these standards

Pollutant	Standard		Government target
	Concentration	Mean measured as	concentration for 2005
Benzene	5 ppb	Annual running	5 ppb
1,3 Butadiene	1 ppb	Annual running	1 ppb
Carbon monoxide	10 ppm	8-hour running	10 ppm
Lead	$0.5\ \mu g/m^3$	Annual	$0.5\mu g/m^3$
Nitrogen dioxide	150 ppb	1 hour	150 ppb
	20 ppb	Annual	20 ppb
Ozone	50 ppb	8-hour running	50 ppb (97th percentile)
Particles PM$_{10}$	$50\mu g/m^3$	24-hour	$50\ \mu g/m^3$ (99th percentile)
Sulphur dioxide	100 ppb	15 minute	100 ppb (99.9th percentile)

Sources: DoE (1997) and subsequent consultations; EPAQS (1994 a–c)

has adopted standards for a variety of toxic air pollutants, all set for the protection of human health, but for no other environmental purpose. Table 3.1 outlines the current standards proposed by EPAQS. The UK government has accepted these as standards, but it has also adopted targets for achievement by the year 2005. Some of these targets are to meet the proposed standard, but some do not go as far. Ozone is a good example of the latter.

EPAQS recommended an ozone standard of 50 ppb as an eight-hour running mean. However, this is widely exceeded across the UK on around 75–80 days per year. In order to achieve such a low concentration, it is estimated that UK emissions of ozone precursors (volatile organic compounds and nitrogen oxides) would need to be cut by between 75 and 95 per cent. The current UK commitment is a 30 per cent reduction. The import of ozone and ozone precursors from Europe makes it difficult to achieve the required standard without international action. The government has therefore stated that it cannot aim to meet the standard in the short term.

These standards can be compared with others set by the EU, the World Health Organisation and the United States (Table 3.2). It can be seen that while there are common approaches across the world, there are also significant differences in the strictness of the standards that are set.

Standards for the natural environment

Standards for the natural environment have not been formerly adopted in the UK National Air Quality Strategy. However, it is perfectly possible to do this. Setting standards for the natural environment can be a difficult exercise, because the objectives of such standards are not always clear. For example, is a standard for

TABLE 3.2 Examples of international air quality standards for human health outside the UK

Source	Sulphur dioxide Period	Concentration ($\mu g/m^3$)	Nitrogen dioxide Period	Concentration ($\mu g/m^3$)	Particulates Period	Concentration ($\mu g/m^3$)
WHO	Annual mean	50	Monthly max	190	Black smoke	
	24 h mean	125	24 h mean	150	Annual mean	50
	1 h mean	350	1 h mean	400	24 h mean	125
	10 min mean	500			Total particulates:	
					Annual mean	60
					24 h mean	120
EU	Annual mean	80 (Smoke > 40)	Limit value:		Black smoke:	
	Annual mean	120 (Smoke < 40)	98th percentile	200	Annual mean	80
	Winter mean	130 (Smoke > 60)	Guide value:		Winter mean	130
	Winter mean	180 (Smoke < 60)	Annual mean	50	98th percentile	250
	98th percentile	250 (Smoke > 150)	98th percentile	135		
	98th percentile	350 (Smoke < 150)				
New Zealand	3 month mean	50	24 h maximum	100	Total particulates:	
	24 h maximum	125			7 day maximum	60
South Africa	Annual mean	80	Annual mean	270	Total particulates:	
	24 h mean	265	24 h mean	540	24 h mean	100
	1 h mean	780	1 h mean	1,080	1 h mean	240
US	Annual mean	80	Annual mean	100	PM_{10}:	
	24 h mean	365			Annual mean	50
					24 h mean	150

sulphur dioxide to be set to protect species present in a location, or to allow the recolonisation of species that have been lost due to past air pollution? Generally, the more ambitious target ought to be adopted. Very few countries have set standards for the natural environment in their statutes. However, for working purposes, the critical levels assessed under the United Nations Economic Commission for Europe (UNECE) provide the basis for operational management in this area. A 'critical level' is defined (CLAG, 1994b) as:

> the concentration [of a pollutant] in the atmosphere above which direct adverse effects on receptors, such as plants, ecosystems or materials, may occur according to present knowledge.

Critical levels (Table 3.3) were identified under the auspices of the UNECE, in order to identify areas where air quality was not good enough to protect the natural environment. The policy environment in which these values were identified did lead

TABLE 3.3 Critical levels of air pollutants

Pollutant	Critical level for vegetation
Ammonia	$8 \, \mu g/m^3$ (annual mean) $23 \, \mu g/m^3$ (monthly mean) $270 \, \mu g/m^3$ (daily mean) $3,300 \, \mu g/m^3$ (1 h mean)
Deposited acidity (rain and cloud)	$1.0 \, mol./m^3 \, H^+$ ions (general vegetation, growing-season mean) $600 \, mol./ha \, H^+$ ions (lichens and bryophytes — annual total deposition) $0.3 \, mol./m^3 \, H^+$ or NH_4^+ ions with $0.15 \, mol./m^3$ non-marine sulphate (forest trees where ground-level cloud is above 10 per cent — annual mean)
Nitrogen dioxide	$30 \, \mu g/m^3$ (annual mean) $95 \, \mu g/m^3$ (4 h mean)
Ozone	Number of ppb hours above 40 ppb for daylight hours: 5,300 (crops) 10,000 (trees)
Sulphur dioxide	$20 \, \mu g/m^3$ (annual mean and half-year mean — vegetation) $10 \, \mu g/m^3$ (annual mean — sensitive lichens)

Source: Ashmore and Wilson (1994)

to some constraints, particularly the need for both pollutant and receptor data to be in a form that was capable of being mapped at a European level.

Critical levels are set to prevent long-term effects on plants. These include both positive and negative effects on growth, as both could have detrimental consequences in natural situations. Critical levels are not set to prevent some measurable physiological responses to pollutants, but to prevent effects that may lead to longer-term damage to species functioning within ecosystems. It could be stated that critical levels are set to prevent impacts that really matter. This could be considered to be equivalent to a definition of 'harm' to ecosystem function, as given in the Environmental Protection Act (1990).

Standards are generally set for operational monitoring periods, e.g. annual average concentrations. However, for many plants, these time periods are not optimal for examining pollutant exposure. For some species, pollutant exposure may be important at certain critical periods, for example, trees are especially susceptible during bud burst, when even very low pollutant concentrations can cause long-lasting effects, but, later in the season, relatively high concentrations cause little effect. From a regulatory viewpoint, such information is difficult to incorporate practically. Thus while one could, in theory, require pollutant levels to be especially low for a short period, this period would vary from year to year, depending on climate, etc. Realistically such a regulatory framework cannot be achieved.

A more common approach is to consider pollutant exposure to plants over periods when they might be more generally sensitive, e.g. the growing season, as is adopted for ozone. The critical levels for nitrogen dioxide and ammonia have not been applied to particular growth periods, but the critical level for sulphur dioxide has been used in this way. Interestingly, it was evident that plants became especially sensitive to sulphur dioxide during cold periods, so that it was recommended that the annual sulphur dioxide critical level of 20 $\mu g/m^3$ should also apply to the winter months for vegetation that overwinters in an active state (e.g. conifers). As production of this pollutant is greatest in the winter, this may mean that the annual average concentration needs to be lower in order to achieve a winter average of 20 $\mu g/m^3$. It is also worth noting that sensitive lichens, with their low critical level of 10 $\mu g/m^3$ are more physiologically active during the winter than at other times.

The use of different averaging periods poses some problems for mappers, in that pollution data have to be recalculated in different ways. However, it poses even greater problems for operators and regulators, as emissions of a particular pollutant from a process may be subject to more than one standard, which needs to be assessed in different ways. However, given the sophistication of current dispersion models, this problem should be able to be overcome, should these standards be incorporated in a regulatory framework.

Local air quality management

The regulation and management of air quality is also implemented at the local level in the UK. Under the 1990 Environmental Protection Act, local authorities regulate smaller air pollution sources than are controlled under IPC. The regulation is only concerned with air pollution, and the task of the authority is to ensure that each process meets nationally-set emission criteria. Under the 1995 Environment Act, this local authority role in Scotland was passed to the Scottish Environmental Protection Agency, although local authorities in England and Wales retained the function.

The 1995 Environment Act also sets a programme for local air quality management, and the UK strategy published in 1997 sets out the particular use of air quality standards at a local level. The Act requires that local authorities (or appropriate groups of authorities, e.g. the London boroughs) undertake regular assessments of air quality in their areas and so assess whether standards are being breached, and, if so, under what circumstances.

Local authorities are required to set up Air Quality Management Areas (AQMAs) 'where it has been shown that the government's targets are not likely to be met solely as a result of national policies'. In each AQMA the local authorities will be required to produce air quality management plans within twelve months, in consultation with the Department of the Environment and with either the Environment Agency or Scottish Environmental Protection Agency, showing how they intend to meet air quality objectives.

However, apart from longer-term planning proposals, e.g. relating to traffic (speed limits, parking, pedestrianisation) or siting of industrial processes, there is some reluctance to allow local authorities to use or have additional emergency powers to deal with air quality problems. For example, it is possible to predict severe smogs when adverse meteorological conditions occur, and a response might be to prevent traffic from entering a town or city where air quality would deteriorate. At present the 1984 Road Traffic Regulations enable local authorities to impose traffic controls on local roads, and the government can impose them on trunk roads and motorways. However, these powers have never been used in response to air quality problems. In the National Air Quality Strategy, the UK government has accepted the principle that such direct management of traffic may be necessary. It remains to be seen what this will mean in practice.

Interpreting BATNEEC in the United Kingdom: regulating individual processes

BATNEEC (best available techniques not entailing excessive cost) has proved one of the more controversial of the components of IPC in the UK. In particular, the interpretation of 'excessive' has caused considerable debate. Some environmentalists claim that it allows a 'cop-out' for the regulator, so that a lack of environmental improvements can always be justified as 'too expensive'. They would prefer a system more in line with Germany, where strict emission standards are set in statute and adherence is mandatory.

However, proponents of BATNEEC argue that it has a number of positive aspects. It forces the operator to look closely at the environmental consequences of the operation. A system of emission standards means that the operator need look no further than the end of the discharge pipe. First, BATNEEC requires the operator to consider whether the discharges are causing harm and, if so, how much and what techniques are necessary to reduce or prevent this. Second, the requirement to look at what is available means that emission standards are not limited to the best achievable at the time of writing the statute. The regulator and operator are both required to keep abreast of technological change. The system does, however, beg the question of how one should weigh up the economic costs of different technological options against the environmental costs of those options.

The Environment Agency has produced a framework for such an analysis (EA, 1997). The framework involves the following steps:

1   A systematic approach to identifying harm in any medium (air, land or water).
2   A method of weighing up impacts to different media and so identifying the best environmental option.
3   Adding an economic analysis to this to assess whether the costs of pollution control options may be excessive in terms of the environmental benefits they bring.

Defining harm

The system adopted is based on environmental quality standards. Some of these are existing statutory standards (e.g. the EC Directive value for air quality for sulphur dioxide or those in the UK National Air Quality Strategy). However, such standards are not available for most substances, and so HMIP has set its own 'standards' or 'environmental assessment levels' (EALs), based on a range of selection methodologies and expert judgement. These have been set for many pollutants for land, air and water. Each EAL is also given an 'action level' at 10 per cent of the EAL concentration. The action level is considered to be the concentration below which environmental effects can be considered negligible.

For any pollutant that is proposed to be released, the concentration in the environment around a particular process is a combination of the concentration of the pollutant released by the process and the existing background levels. Together they form the 'predicted environmental concentration'(PEC). For existing processes, monitoring around a plant will measure the PEC. For a new process, monitoring will identify background levels, and dispersion modelling will be necessary to identify the contribution from the plant.

Under the assessment methodology, whereby a pollutant release would result in a PEC greater than the EAL, this would be considered as unacceptable and alternatives would need to be considered (including not operating the process). If the PEC breaches the action level, then BATNEEC should be used to prevent releases, or minimise or render them harmless. If the PEC is below the action level, no action is necessary by the operator or the regulator.

Defining the best environmental option

In order to estimate the relative contribution of the process to pollutant problems in the different environmental media, the Environment Agency has developed an 'integrated environmental index'. For each pollutant, for each medium, the long-term concentration of the pollutant is compared to the EAL:

plant contribution

————————————

EAL

This estimates the proportional emissions of the plant towards a concentration that would breach an EAL. The sum of all such calculations can be made for all air pollutants, water pollutants and land contaminants. The overall sum is the integrated environmental index.

Different plant operational options (e.g. changing discharges from one medium to another, or adopting new technologies) can be assessed in exactly the same manner. The option that produces the lowest integrated environmental index is defined as the best environmental option for the local impacts of pollutant releases.

However, the emissions from a process can have environmental consequences other than their effects on local environmental quality standards. For example, emissions of chlorofluorocarbons can lead to reduction in stratospheric ozone and contribute to global warming. On a regional scale, emissions of volatile organic compounds can contribute to tropospheric ozone production and hence to adverse air quality. The overall BPEO assessment has to weigh up these factors in addition to those affecting local environmental quality. This approach has been criticised for trying to compare 'apples and oranges'. However, it is very difficult to identify other methods that can quantitatively compare the relative consequences of breaching environmental standards in air or water.

Incorporating economic costs

The HMIP assessment methodology uses a simple test for 'excessive' cost. The economic costs of various options can be compared to the environmental benefits that they bring. Where there is a significant rise in cost without a significant improvement in environmental benefits, then this 'break-point' may be where further costs are no longer justified. However, this does assume that such break-points will be clearly identifiable. This part of the methodology, particular, has yet to be fully tested and is likely to change.

Air pollution regulation by the European Union

The European Community (now the European Union) has passed a series of directives regulating aspects of air pollution. Currently, this area of environmental protection is being actively developed by the EU, and the following discussion should be considered in the light of new directives. These cover a series of different areas.

Air quality standards

In 1980 the EC passed a directive introducing standards for smoke, sulphur dioxide and, subsequently, for nitrogen dioxide and lead in the atmosphere. These standards included limit values, which should not be breached, and guide values, which ought to be met. These values included both long-term mean concentrations and peak concentrations. The standards were set only to protect human health.

In 1992 the EC adopted a directive on ozone pollution. This set standards for the protection of health and for vegetation. It also introduced a value for health which would result in the government of a member state being required to issue a public health warning.

The European Commission is currently proposing a framework for the implementation of standards throughout member states. These would be legally binding and set a series of clear stepwise targets, with fixed time limits towards meeting each standard.

Air quality monitoring

A 1982 directive requires the establishment of air quality monitoring in member states and for the exchange of the information collected. The EU has also legislated on some specific impacts monitoring, for example a 1978 directive on the requirement for a blood screening programme to assess levels of lead. However, monitoring has now been encapsulated within the recent Ambient Air Quality Assessment and Management Directive. This provides an overall framework whereby 'daughter directives' will be agreed to set air quality limit values for individual pollutants as well as requirements by members states to monitor air quality and draw up action plans for dealing with air quality problems. In the UK the measures adopted in the 1995 Environment Act should meet the requirements of the Directive. The most important development in this area has been the establishment of the European Environment Agency, which will collate data from member states and monitor progress towards the achievement of environmental quality standards.

Emission limits

Emission standards have been set under a series of directives to ensure compliance across the Union. These include requirements for a range of motor vehicle types and municipal waste incinerators. It is likely that controls on VOC emissions will also be required in the near future.

Fuel specifications

Directives have also been passed to require improvements to the quality of fuels used for combustion, and hence improve emissions without technological changes to industrial processes or motor vehicles. Examples include the sulphur content of fuel oils, and the lead and benzene content of petrol. Further controls are still being developed.

Limits on total country emissions

The EU has focused on trans-boundary and wider international air pollution issues. The 1988 Large Combustion Plant Directive set specific limits for each member state for emissions of sulphur dioxide and nitrogen dioxide from large fossil fuel processes (50 MW and above), with target dates for compliance. For example, the UK was committed to reductions in sulphur dioxide emissions from a 1980 baseline of 20, 40 and 60 per cent by 1993, 1998 and 2003 respectively, and for reductions in nitrogen dioxide of 15 and 30 per cent by 1993 and 1998 respectively. The Directive is currently being revised and will adopt an 'effects-based' approach already used under the UNECE (see p. 77).

The EU has also played a part in phasing out the use of chlorofluorocarbons within the Union, under the Montreal Protocol of 1988. The Council of Ministers agreed a ratification of the Protocol and a timetable for compliance for all member states to achieve.

Market mechanisms

The use of market mechanisms by the EU is a relatively new consideration. While it is favoured philosophically by a number of member states in mirroring changes to domestic regulatory systems (e.g. the UK), it does fall foul of a distrust by member states of direct EU control of taxation measures. The only proposal so far has been to suggest a tax on fuels emitting carbon dioxide, as a means of reducing greenhouse gas emissions. However, the early support for the proposal has declined and the implementation of such a measure is uncertain. An alternative market mechanism is that of a permit trading system. While no firm proposals currently exist, the potential use of such a system for further control of sulphur dioxide emissions is being considered.

An overall regulatory framework

In September 1996 the EU agreed a Directive on Integrated Pollution Prevention and Control (IPPC). This aims to regulate pollution from the following categories of process:

- energy industries;
- production and processing of metals;
- mineral industry;
- chemical industry;
- other activities (including intensive animal housing).

Each member state is required to introduce legislation and regulations by 20 October 1999 to implement the Directive. All existing processes will have to be authorised under the IPPC conditions by 30 October 2004, and new processes require prior authorisation as soon as domestic legislation is in place. The basic aim of the Directive is to ensure that the competent authorities in the member states regulate the processes to ensure that:

- all appropriate preventative measures are taken against pollution;
- no significant pollution is caused;
- waste production is avoided;
- energy is used efficiently;
- measures are taken upon completion of process activities to prevent subsequent pollution.

The approach is to achieve BPEO and to consider this in a wider context than the simple process operation. For example, in the UK IPC seeks to prevent or minimise emissions of *prescribed substances*, and to prevent harm from other released substances. However, IPC does not control the release of harmless substances, and this may actually have significant resource implications. IPPC will consider the whole issue of resource use, waste minimisation and energy efficiency and will consider all activities at the location of a process, not just the narrow confines of a defined process type.

'The emphasis in IPPC is more closely linked to BAT (best available techniques) than UK legislation. However, the pressure for common standards across the EU (which is supported by Germany and the Scandinavian countries) has had added to it a discretion to allow member states to consider local environmental conditions in setting emission limits, i.e. these could be relaxed where environmental quality standards are not likely to be breached. This partially meets the concerns of South European states. Similarly, if compliance with BAT standards would still lead to breaches of standards, additional measures could be taken. In practice, therefore, the application of standards may not be dissimilar to BATNEEC in the UK.

The adoption of IPPC will mean a number of changes in UK pollution regulation. A number of processes (e.g. food and drink manufacturing, small metal industries and solvent users) are currently under LAAPC or are even unregulated. These will be brought under IPPC. The UK can decide to regulate these through the Environment Agency, as with IPC, or coordinate their regulation through existing bodies. Of considerable interest will be the regulation of intensive livestock units. These are major sources of ammonia (an important contributor to acid rain), methane (a major greenhouse gas) and of odour nuisances. Currently unregulated for aerial emissions, the adoption of IPPC will extend the authority of the air pollution regulators into the agricultural sector, an unfamiliar experience in the UK.

Currently, the European Commission is also considering a new framework directive on a strategy for regulation of emissions from small industrial processes. This would include provisions for emission limits, regulatory administration, monitoring and data exchange. Together with IPPC, the aim would be that the two directives would encompass 'the totality of European industries responsible for dangerous and long-range emissions'. However, it would still leave open the control of *diffuse* pollutant sources.

CASE STUDY: AIR POLLUTION CONTROL IN POLAND

The former Communist states of Eastern Europe became notorious for the pollution problems caused by their industrial activity. In effect the lower efficiency of such industries in comparison to the West was partly compensated for by very low environmental standards and, therefore, reduced costs in this area. Many of these countries have embarked on ambitious clean-up programmes. A good example of this is Poland.

Since the fall of its Communist government, Poland has embarked on a range of environmental measures to reduce air pollution. The concentrations of many pollutants have been above those necessary to protect human health and a three-stage programme has been adopted by the Ministry of Environmental Protection, Natural Resources and Forestry. This programme has short-term, medium-term and long-term actions.

The short-term programme is aimed at dealing with significant immediate problems. It includes:

◆ emission controls on motor vehicles (including use of lead-free petrol and catalytic converters);
◆ measures aimed at the major industrial polluters;
◆ improvements to the quality of coal used in combustion processes (e.g. by coal washing);
◆ a monitoring programme for areas with poor air quality and a warning system for pollution episodes. This is aided by a loan from the World Bank Environmental Management Project.

The medium-term programme is based on achieving environmental standards. In preparation for future membership, the standards adopted are those of the EU and the aim is that these are achieved by the year 2000. This requires reductions (from a 1980 baseline) of sulphur dioxide by 30 per cent, nitrogen oxides by 10 per cent and particulates by 50 per cent.

The long-term programme is much less specific and seeks to adopt sustainable development principles by the year 2020.

The pattern of pollution emissions in Eastern Europe has not been even. Industry has been highly concentrated around sources of raw materials. The Polish strategy recognises this and includes a regional dimension to pollution control.

It works by the national government setting environmental standards and general emission standards. However, the Ecological Department in each Province (which authorises industrial emissions) can set stricter local emission standards if there are particular local problems.

Poland has adopted widespread emission charges as an economic instrument in achieving pollution reduction. Few European countries use this method, but Poland has set rates for sixty-three pollutants. Each company pays the same for a given pollutant and there is therefore a strong incentive to reduce emissions below those authorised. Emissions above authorisation generally result in fines at about ten times the emission charge. Provincial governments do not have powers to set different rates of charge. The government of Katowice attempted this in the early 1990s, by doubling the charge to reduce their extremely bad local pollution problems. However, the Polish supreme court ruled that this was unconstitutional. The income from pollution charges is channelled to environmental funds, with proportions earmarked for national and local projects. Payments from these funds may be made to aid industry to improve its performance.

Much of the reduction in air pollution in Poland is due to the closure of inefficient industry as the free-market has penetrated the post-Communist system. Very often those plants which were closed were particularly polluting. Some of the reduction has, however, taken the form of reduced production (e.g. electricity generation) and it is important that, as the economy improves, the increased activity in such plants is achieved while meeting environmental standards. The power sector has been adopting flue-gas cleaning technology and clean coal technology to meet this.

The government has recognised the value of public interest in environmental matters. This is in stark contrast to the secrecy prevalent under the Communist regime. The main *NGO* is the Polish Ecology Club, which has worked closely with government and has campaigned vigorously at a local level. Of particular importance has been the need to inform the public of environmental issues. The Ministry of the Environment maintains an inventory of major polluters and this provides a 'league table' for public scrutiny. The pollution monitoring programme also has specific measures to inform the public of air quality changes and problems.

Poland has, therefore, embarked on a structured programme to clean up its industry. This is focused on environmental goals, regional variation and the need to meet economic and environmental standards prior to EU membership. Some help has be received from the international community, but much of the progress has been achieved by strong determination from the people themselves.

The United Nations Economic Commission for Europe (UNECE)

Many air pollution problems extend beyond the European Union. For example, many important sources of acid rain occur in Eastern Europe, and some important recipients of acid rain (e.g. Norway) are outside the EU. A wider forum is therefore necessary to achieve international agreements on the control of such pollutants. The

United Nations has set up an Economic Commission for Europe. This includes all states within Europe, from Ireland to the Urals, as well as Canada and the United States. In 1979 the UNECE agreed a Convention on Long-Range Trans-boundary Air Pollution, under which a series of protocols have been agreed. This was put into force in 1983. The early protocols were related to monitoring and mapping of pollutants. However, the later protocols have set agreements on stepwise reductions in the pollutants themselves. Agreed protocols are as follows:

1   'Protocol on the Long-Term Financing of the Cooperative Programme for Monitoring and Evaluation of Long-Range Transmission of Air Pollutants in Europe (EMEP).' Agreed on 28 September 1984. Ratified by the UK.
2   'Protocol on the Reduction of Sulphur Emissions or Their Trans-boundary Fluxes by At Least 30 per cent.' Also known as the '30 per cent club', requiring a 30 per cent cut in total sulphur dioxide emissions from 1980 levels by 1993. Agreed on 8 July 1985. The UK did not sign, received much adverse reaction and then met the target in any case.
3   'Protocol concerning the Control of Emissions of Nitrogen Oxides or their Trans-boundary Fluxes.' Requires a reduction in NO_x emissions to 1987 levels by 1994. Agreed on 31 October 1988. Ratified by UK and UK target has been achieved.
4   'Protocol concerning the control of emissions of Volatile Organic Compounds (VOCs) or their trans-boundary fluxes.' Requires a 30 per cent reduction in VOCs from 1988 levels by 1999. Agreed on 18 November 1991. Ratified by UK, but has still not come into force.
5   'Second Protocol on the Further Reduction of Sulphur Emissions.' Requires specific reduction in sulphur emissions from different countries based on critical loads analysis (see p. 77). Agreed on 14 June 1994. UK has not yet ratified it and it has yet to come into force.
6   Currently under negotiation are a revised protocol for nitrogen oxides which will also take account of ozone generation (and thus VOCs again), general acidification (possibly re-examining sulphur) and eutrophication (including ammonia) and protocols to reduce emissions of heavy metals and persistent organic toxins.

The earlier protocols above have been based on an arbitrary reduction in emissions by all parties. However, it is clear from monitoring and modelling work that emissions from one part of Europe do not have equally damaging consequences in other parts of Europe. A far better approach is to target those emissions that cause the most damage. This is the so-called '*effects-based approach*' and was the basis for the revision of the sulphur protocol signed in 1994. This took account of the amount of acidifying pollutants that the ecosystems of Europe could withstand, i.e. their critical load (see p. 77) and combined this with models of pollutant dispersion from each country. The models were optimised to target the most important sulphur sources, in order to achieve a given environmental benefit. The current revision of the nitrogen oxide protocol is also adopting an effects-based approach.

The critical loads approach: assessing national and international policies

The basis of the critical loads approach is that there is a threshold to the effect of a pollutant in an ecosystem (Bull, 1991, 1992). The common definition has been that adopted at a UNECE workshop (Nilsson and Grennfelt, 1988):

> a quantitative estimate of an exposure to one or more pollutants below which significant harmful effects on specified sensitive elements of the environment do not occur according to present knowledge.

If pollutant exposure is greater than the critical load, then the critical load is considered to be 'exceeded'. The definition has a number of important components to it.

First, it assumes a threshold. For some pollutants this is reasonable. Hydrogen ions, for example, occur naturally in rainfall and thus ecosystems experience a natural background level. However, for non-natural pollutants (e.g. dioxins or radionuclides) such a concept may not be appropriate.

Second, the requirement is for a 'quantitative' estimate. This is necessary in order to use the result in assessing the emission reductions necessary to prevent the critical load being exceeded. Producing a quantitative estimate does focus the minds of researchers in refining data, but the provision of 'best estimates' can lead to unfounded confidence by policy-makers, who may rely on these numbers.

Third, the definition refers to exposure. This is commonly considered in two ways, either as deposition of a pollutant (e.g. sulphur, acid or nitrogen) or as a gaseous concentration to which an organism is exposed. The former retains the concept of critical 'load', but the latter is better known as the critical 'level'. The impacts under assessment are described as 'significant'. Trivial pollutant responses are not considered to be important, but those affecting plant growth, crop yield or ecosystem structure are.

Fourth, the critical load is set to protect 'sensitive' components of the environment. Very often such components are of the most importance, e.g. they are rare species in particular habitats. Strategies to achieve environmental protection must include these.

Finally, the definition adds 'according to present knowledge'. The approach is based on scientific understanding of the emission, transport, deposition and impact of the pollutants. This knowledge base is always expanding and any numbers generated are open to revision.

Current use of critical loads in policy decision-making has been based on chemical responses in environmental receptors, i.e. soils and freshwaters. Soils critical loads maps are currently based on the estimates of the weathering rates of the component minerals, i.e. how quickly they release the base cations that can buffer any acidic inputs to deposition. This allows the use of existing detailed soil maps and, given uncertainties in modelling, provides readily quantifiable estimates. However, it is generally difficult to relate biological responses of organisms in soils to the criteria set for critical loads. Although in taking a precautionary approach, it is generally assumed that long-term change in soil chemistry is unacceptable. Similarly, fresh-water critical loads are based on achieving a given acid neutralising capacity (ANC). The chosen ANC can be set to particular biological requirements, e.g. an ANC = 0 would protect salmonid fisheries. Future development of the critical loads approach will link more explicitly the biological response to critical loads values, which is especially necessary for the issue of nitrogen to be addressed.

CASE STUDY: COMBINING CRITICAL LOADS AND DEPOSITION DATA TO ASSESS POLICY OPTIONS

Models exist which estimate the deposition of pollutants. These can be based on single countries or are Europe-wide. It is necessary to undertake such modelling, as while current monitoring of deposition can describe impacts today, policy-makers wish to consider the options for tomorrow. They need to ask questions such as, how much deposition reduction is necessary to prevent exceedence of critical loads? What is the effect of preventing emissions from a particular power station? Would a particular reduction cause more environmental benefits if controls were introduced in the UK or in Poland?

Some of the most detailed policy analysis using critical loads assessing impacts on the environment has been undertaken within the UK. An analysis by the GB conservation agencies (Farmer, 1993a; Farmer and Bareham, 1993) assessed where the critical loads for sulphur deposition in areas of Britain containing statutory nature conservation sites were exceeded and considered whether these might be susceptible to acidification. Modelling allowed an assessment of different future policy options, i.e. progressive 60, 70, 80 and 90 per cent reductions of sulphur dioxide emissions from a 1980 baseline. A 60 per cent reduction was the then limit under the EU Large Combustion Plant Directive for the UK for the year 2003, and 70 per cent and 80 per cent are new commitments under the 1994 revised UNECE Sulphur Protocol for the UK for 2005 and 2010 respectively. Table 3.4 presents the results of the analysis.

One conclusion to be drawn is that it is evident that equal reductions in emission do not provide equal protection of sites. For example, the reduction from 70 to 80 per cent protects about twice the site area in England and Wales as the reduction from 60 to 70 per cent. Thus such analysis helps to ensure that commitments do achieve significant environmental benefits.

However, the analytical procedure does reveal some drawbacks. During the sulphur protocol negotiations there was considerable debate surrounding the issue of target dates and many countries not only adopted stricter reductions than the UK, but agreed earlier dates for achievement. However, the critical loads approach, as used here, does not allow an understanding of the consequences of, for example, the UK achieving an 80 per cent reduction by 2005 instead of 2010. The models are 'steady-state' rather than 'dynamic'.

TABLE 3.4 Area and number of statutory nature conservation sites in Britain that remain at risk from acidification under four sulphur emission reduction projections, from a 1980 baseline

		Area (ha) of sites at risk			
		60 % reduction	70 % reduction	80 % reduction	90 % reduction
England	Area	211,919	187,933	126,324	79,625
	Number	600	455	250	94
Scotland	Area	148,614	—	52,443	—
	Number	205	—	69	—
Wales	Area	109,501	862	46,609	28,042
	Number	218	151	86	60

Monitoring air pollution

Monitoring is now well established as a necessary component of any strategy to manage air pollutants. There are a number of different reasons why air pollutants can be monitored. Broadly, these fall into two areas, i.e. understanding the nature and behaviour of air pollutants, and targeting and assessing the effectiveness of emission control measures. Particular monitoring issues include the following.

Pollutant distribution

There are now a number of well-established monitoring networks around the world. These may measure a wide range of gaseous and wet deposited pollutants. Those pollutants which require simple monitoring techniques tend to result in more comprehensive networks, whereas those requiring expensive and sophisticated equipment are more often monitored less frequently due to limitations of money, experienced personnel and, occasionally, problems of equipment security.

Figure 3.1 shows the distribution in the UK of the primary and secondary acid rain monitoring network. This contains primary sites which monitor a wider range of pollutants more frequently than the secondary sites, allowing flexibility in assessing either details in pollutant distribution or in the behaviour of particular pollutants.

Regular monitoring also allows for the examination of trends in pollutant concentrations over time, thus setting current pollution management objectives in a historic framework.

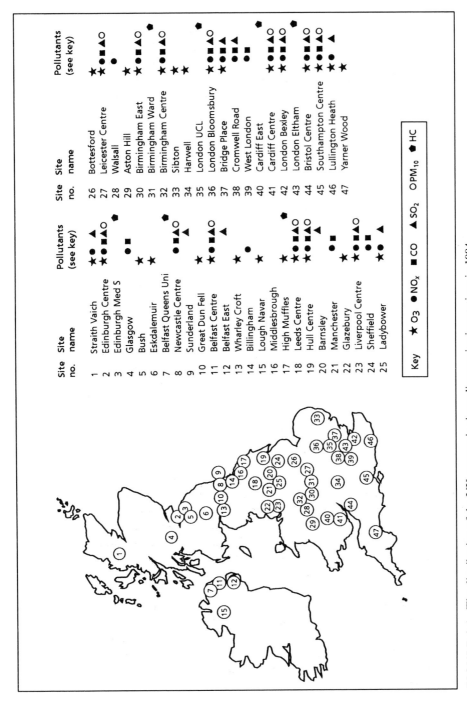

FIGURE 3.1 The distribution of the UK automatic air quality monitoring stations in 1994
Source: Bower (1996).
Reproduced from *Urban Air Quality and Public Health*, with kind permission of ENSIS Publishing Ltd.

Pollutant behaviour

National networks and monitoring around individual processes may be undertaken to further the understanding of pollutant behaviour. Different pollutants, for example, do not deposit at the same rate, and monitoring can provide information on this. Pollutant interaction is often difficult to predict. Thus monitoring has helped our understanding of how, for example, nitrogen oxide and volatile organic compound pollutants interact to form ozone in the presence of sunlight.

Pollutant emissions

It is possible to estimate emissions from some industrial processes by examining details of process operation. For example, reasonable estimates can be made of overall sulphur dioxide emissions by examining changes in the sulphur content of the fuels being burnt. Indeed, in proposing new processes that result in pollution, calculations of potential emissions are the only means of assessing the likely impact. However, it is important that for most processes there is verification of emission estimates. This requires monitoring in the stacks through which the pollutants are emitted.

Stack monitoring is particularly important where pollutant concentrations may vary widely. One example would be the incineration of waste. Waste combustion requires the maintenance of very high temperatures to ensure complete destruction of the waste and to, for example, prevent the formation of dioxins from the incineration of plastics. However, the waste itself provides much of the fuel and, as this is very variable, the temperatures may vary considerably. While modern incinerators supplement the waste with gas burners to ensure temperature maintenance, it is important to monitor the flue gases to ensure that changes in conditions do not lead to unacceptable pollutant emissions.

Stack monitoring is not always possible. There is considerable concern at present regarding the overall emissions of the volatile organic compounds that act as precursors for ozone formation. A large source of these are the solvents used in industry. However, rather than being emitted from one or a few chimneys, these tend to arise from many points in an industrial plant, often from leaks, etc. These are known as *fugitive emissions*. It is possible to estimate loss by keeping a careful balance sheet of solvent use and recovery and assuming the difference is lost to fugitive emissions. However, monitoring the environment around the process can also provide information on these emissions.

Obviously, monitoring stack emissions does enable an operator to judge whether pollution abatement equipment is functioning adequately. Monitoring equipment would, for example, be usefully used to assess when particulate control techniques (e.g. bag filters) need to be replaced. A more familiar example, however, is that of the emissions check to motor vehicles as part of the annual Ministry of Transport test (MOT) in the UK.

Monitoring to confirm model predictions

When proposals to operate an industrial process are made, or when planning decisions are made concerning transport management, models are used to predict pollutant concentrations in the environment. If these are considered acceptable, the process is authorised or the plan accepted. However, some form of monitoring is often required in order to ensure that the modelling results were correct, and therefore to trigger remedial action if necessary.

In the UK a national monitoring network has been established for a range of purposes, e.g. acid rain monitoring or urban air quality. Additionally, many local authorities monitor air quality, as do major industrial companies. Table 3.5 indicates the number of sites for different pollutants planned for the end of 1996 in the UK that are within the national network.

TABLE 3.5 Number of sites for different pollutant types in the UK National Automatic Monitoring Network by the end of 1996

Pollutant	Number of sites
Carbon monoxide	57
PM_{10}	50
Nitrogen dioxide	84
Sulphur dioxide	65
Benzene and 1,3-butadiene	12
Ozone	74

It is important to note that different monitoring stations are not always equivalent, even when they monitor the same pollutants. Some monitoring may be undertaken with high-tech automatic analysers, which provide accurate information for very brief time periods. These are very expensive to install, and require regular maintenance and calibration. At the other end of the scale, there are simple passive samples (a plastic tube with a chemical absorbant at one end), which costs only a few pounds. Each has advantages and disadvantages, and a pollution manager should first identify what the purpose of any monitoring regime should be and then seek the most cost-effective monitoring techniques to achieve this. The UK National Monitoring Network incorporates all types of samplers, which provide a range of different information. Bower (1996) has summarised the pros and cons of the different sampling techniques (Table 3.6).

Modelling air pollution dispersion

A very wide range of effective models now exist for assessing dispersion of pollutants from point and linear sources. As many of these have been created privately or

TABLE 3.6 Advantages and disadvantages of different air quality monitoring techniques

Method	Advantages	Disadvantages	Cost
Passive sampler, e.g. nitrogen dioxide diffusion tube	Very cheap Very simple Easily hidden from view No electricity requirement (good in remote areas)	Unproven for some pollutants Lab. analysis required (slow to obtain data) Minimum averaging period is a week	£5–£50 per sample
Active sampler, e.g. sulphur dioxide 'bubbler'	Cheap Simple Reliable	Labour intensive Lab. analysis required (slow to obtain data) Minimum averaging period is a day	£1,000–2,000 per unit
Automatic analyser, e.g. ozone monitor	High proven performance Hourly data	Complex Expensive Requires skilled staff	£10,000 per unit (with possible high recurrent costs)

Source: Modified from Bower (1996)

modified for particular companies or applications, it is inappropriate to review particular models here, as many non-commercially available models may be equally or even more applicable to a given environmental assessment. A number of reviews do exist, however, of the different types of model (e.g. Zannetti, 1992). The information obtained from a model is, however, only as good as the information put into it and the precision of the questions being asked.

Dispersion modelling

The dispersion of a pollutant from its source depends on the combination of three factors: the nature of the effluent, the nature of the pollutant and the behaviour of the atmosphere. Modelling combines information on all of these factors to predict the rate of spread of the pollutant and hence its rate of dilution in the environment.

It is important always to bear in mind the principles upon which a particular model are based. For example, a simple Gaussian dispersion model provides a estimation of dispersion by describing the pollutant cone from the stack in terms of the standard deviation of concentration from the centre line of that cone. Wind direction will define the trajectory of the cone, and variations in wind speed will define the shape of that cone, i.e. the value of the standard deviation. Different weather patterns are used to assess the shape of the cone to produce points at which the cone will come to ground and thus where maximum ground-level concentrations of pollutants will occur. This can be further refined, with constraints of topography being imposed on the model.

Models are not perfect predictors of how the dispersal of a pollutant will behave in reality. It is important to understand the limitations of any model being used. Many Gaussian models, for example, have difficulties in predicting effective stack height under turbulent weather conditions. If possible, an environmental statement using dispersion modelling ought to indicate the limitations and uncertainties contained in that model. This is particularly important for the non-technical audience, who will not be familiar with these techniques.

As variations in a model run are tried, the operator must consider how the model is processing the information. This is particularly important with the more complex models, in order to ensure that results make sense.

A number of factors can be incorporated into dispersion models:

1   Nature of the pollutant. Emissions may be solid (e.g. particulates) or gases. The former will show sedimentary behaviour, while the latter will diffuse.

2   Effluent characters. The efflux velocity of the effluent and its buoyancy are necessary to calculate the effective height from which dispersion can be said to occur.

3   Emission rate (g pollutant per second); alternatives to the model run may assess the benefits of different fuel types, gas cleaning, etc.

4   Source dimensions, i.e. stack height and width; alternatives to a particular height are easily assessed.

5   Wind speed; alternatives can consider averages and worst cases.

6   Wind direction; alternatives may assess prevailing winds or incorporate full data from a *wind rose*.

7   Atmospheric stability; alternatives should consider different weather patterns (known as Pasquill stability categories), but should do this in a sensible way, so that full information is provided on the usual conditions as well as on the various extreme events, together with an estimation of the frequency of their duration.

8   Topography; effects of nearby buildings, hills, valleys, etc. are all necessary in order to identify how the *plume* may behave before reaching particular sensitive locations. The importance of incorporating topographic data into a model is more important with a topography that is more variable.

9   Existing pollution climate. Variations in the current pollution climate need to be incorporated in order to assess the relative contribution from the process and to identify whether peak concentrations caused by the process coincide with peaks for that pollutant in the environment as a whole.

Deposition modelling

Processes that emit significant quantities of pollutants may need to address not only the effects of their operation on local ground-level concentrations, but also need to look at the impacts of deposition. The long-term deposition of some pollutants will affect the character of the soil and vegetation, thus changing the nature of a given habitat. This may occur even if ground-level concentrations are below those known to cause direct toxic effects on health or vegetation.

Pollutants likely to be of most importance in considering deposition include: the acid gases (for estimates of total acid input), nitrogen oxides and ammonia (for estimations of total nitrogen deposition), dust and fluoride (both as hydrogen fluoride and as particulate fluoride) and particulate metals. Some of these pollutants are well understood in terms of deposition, especially the behaviour of particulates. For others, the science is rapidly advancing and it would be inadvisable to provide figures for deposition rates here, when their accuracy may soon be called into question.

Deposition can occur through a number of pathways. Gases like ammonia will deposit directly on to wet surfaces, e.g. leaves and also through the stomatal pores in the leaves. Thus the amount deposited will depend upon the surface area of the vegetation and on its type. Trees, for example, provide a very large area for deposition. Some plants (e.g. lichens and mosses) have no resistance to gas uptake, i.e. they have no impervious cuticle on their leaf surfaces, and so deposition will occur more readily.

For most purposes, deposition effects can be calculated using data on annual average ground-level concentrations. However, it is evident that for a given concentration a range of environmental factors will need to be assessed:

1   General climate; wetter regions will need a greater consideration of wet deposition over dry deposition.
2   Vegetation characteristics; the type of vegetation and its composition may need to be assessed in order to identify accurate deposition rates for use in models.
3   Pollution climate; the effect of the process will need to be considered in comparison with background deposition.

Basic dispersion models

A full discussion on which dispersion models are appropriate under which conditions is given in EAC (1996), and is beyond the scope of this volume. However, the most basic form of dispersion model is based on the Gaussian plume dispersion model, and some appreciation of how this works would aid the reader. The model assumes that the plume increases in size in a perfect cone shape. The concentration is highest in the middle and is normally distributed towards the edge. It is therefore possible to estimate statistically the concentration in any part of the cone. The standard deviation of the concentration increases with increasing distance from the source as the plume expands.

The plume is centred around the 'effective' height of the chimney. This is not just the actual height, but the additional height given by the buoyancy of the hot gases. The concentration obviously depends upon the strength of the source (given in g/s) and dispersal is related to wind speed.

The overall Gaussian plume-dispersal equation is quite complex. However, we are not usually interested in how pollutants are dispersing everywhere, but on their effect on ground-level concentration. The equation can be simplified as follows:

$$C = \frac{Q}{\pi \sigma_y \sigma_z u} \left[\exp \frac{-1}{2} \left(\frac{He}{\sigma_z} \right)^2 \right]$$

where: C is the ground-level concentration (g/m^3); Q is emission strength (g/s); σ_y and σ_z are the standard deviations (horizontal and vertical); u is the wind speed (m/s); and He is the effective chimney height (m).

The equation does not include the distance from the source, but this is contained in the two factors for the standard deviation.

The two terms for standard deviations vary, depending on the physical characteristics of the atmosphere. The combination of wind speeds, sunlight, etc. is indicative of rapid dispersal or inversions that may increase local concentrations. The different types of atmospheric stability have been classified into different 'Pasquill Stability Categories'. These are given letters, with the most common for the UK being category 'D', which changes the values of the standard deviations. For class 'D' the values are determined for different distances from the formulae:

$$\sigma_y = 0.13 \, (x^{0.903}) \quad \sigma_z = 0.20 \, (x^{0.76})$$

where x is the distance from the source in metres.

This model can also be modified to estimate dispersion from linear sources such as roads. In this case the equation is:

$$C = \sqrt{\frac{2}{\pi} \left(\frac{qN}{u\sigma_z} \right)}$$

where q is the rate of emission (g/m/vehicle) and N is the traffic flow rate (vehicles/second).

There are now a wide range of models in use by industry and environmental consultancies, all more sophisticated than that described above. It is important to ensure that in each case the strengths and weaknesses of each model are understood. Some models may be very good for particular meteorological conditions or are particularly weak on assessing peak *ground-level concentrations*. For example, a recent analysis of common UK dispersion models with ambient air quality data (EA, 1996c) has shown significant discrepancies, which may affect the way that regulatory decisions may be made.

Techniques to reduce air pollution

Reducing air pollutants from industry

Industry, in its broadest sense, has been and still is a major contributor to air pollution. However, the management of these sources of pollution has not always been undertaken in the most efficient manner. The following two points should be considered in any current management strategy.

1   A holistic view of pollutant emissions must be taken. Thus strategies to reduce air pollution must not lead, for example, to greater water pollution.
2   The whole process operation must be examined. Far too often environmental managers and engineers have sought answers to what to do with pollutants once they are emitted ('end-of-pipe' solutions), rather than seeking to prevent their formation. Pollutants may actually represent losses of valuable material (e.g. solvents), and measures to prevent their loss may actually save money.

Changing the nature of the fuel

Where pollutant emissions are due to the type of fuel being used for combustion, modifying the fuel can have significant effects on emissions. For example, when the major power stations in England and Wales were privatised in the late 1980s, there was an agreement that *flue-gas* cleaning would be undertaken on a large proportion of them in order to reduce sulphur emissions. However, the scale of such cleaning was deemed unnecessary by the generators, as they could meet the environmental objectives without spending on emission controls, first by the selective burning of low-sulphur coal and second by the construction of gas-fired power stations, which emit negligible quantities of sulphur.

It is also possible to use traditional fuel sources and treat them prior to combustion to remove substances, such as sulphur, which may cause air pollution. A number of techniques have been examined for the removal of sulphur from coal, for example. Physical cleaning of coal is now well tested. This is achieved by crushing the coal into fine particles and separating the sulphur in the form of pyrites. This can remove about 50 per cent of the sulphur present. However, the process does require additional energy consumption and produces significant quantities of solid and aquatic waste.

Current state-of-the-art treatment for coal (and oil) prior to combustion is integrated gasification combined cycle (IGCC). This process involves the conversion of solid and liquid fuels to a gas. Gases are much more easily cleansed of sulphur and particulates. Sulphur removal is about 99 per cent efficient. The process also produces little secondary waste. Indeed, it can result in the sale of by-products of elemental sulphur and hydrogen.

Changing process conditions to reduce pollutant production

Some pollutants are produced during the process itself. Examples include the production of nitrogen oxides during combustion, or dioxins during incineration. Alterations to the way that processes are operated can significantly reduce the creation of these pollutants.

The amount of nitrogen oxides produced can be reduced in various ways. These oxides are produced by the oxidation of atmospheric nitrogen and oxygen in the combustion chamber. It is possible to achieve a 20–30 per cent reduction in nitrogen oxide production with careful control of the amount of air admitted into the combustion chamber. However, care has to be taken, as an air content that is too low may result in the creation of enhanced carbon monoxide levels. An increasingly used technique for nitrogen oxide reduction is the use of low NO_x burners. These carefully control the fuel/air mix and have been developed for coal, oil and gas use. They can achieve a 30–50 per cent reduction in the production of nitrogen oxides.

One of the major concerns with waste incineration is the production of dioxins in the emissions. Dioxins are readily destroyed if the combustion temperatures are kept sufficiently high (not less than 1,000°C). For many combustion processes this is easily achieved, but waste incineration, by its very nature, tends to involve rapid variations in the types of 'fuel' (i.e. waste) being burnt, and hence large changes in its calorific value. To ensure maintenance of high temperatures, accurate monitoring of temperatures within or immediately following the combustion chamber are required, and it is likely that additional fuel (usually natural gas) may need to be supplied to maintain the temperature if combustion of the waste cannot do this. Gas-firing is also required during the start-up of an incinerator to ensure correct temperatures are achieved before any waste is burnt.

Cleaning the flue gases

If it is not possible to prevent the production of pollutants, then it is likely to be necessary to prevent their release into the environment by cleaning the exhaust gases.

Almost all processes have some form of gas cleaning, generally to remove particulates (ICE, 1985). A range of techniques are available, depending on the size of the process and quantities of pollutant. For small processes, particulates can be collected from the waste stream, using filter bags through which particles above a given size cannot pass. These are efficient, but do become blocked and need regular replacement. The most common technique for particulate removal is the use of cyclones. There are a wide variety of these. In effect cyclones can be viewed as containers that form a large expansion (increased diameter) within the exhaust pipes, etc., through which the waste gases pass. By increasing the size, the velocity of the gases decreases and particles can settle out. With correct positioning of inlets and outlets to cyclones, they can form efficient particulate traps. If high levels of particulate retention are required, more than one cyclone can be used in a progressive series. Because cyclones operate by means of physical deposition of particles from the waste stream, they operate more efficiently for larger particle sizes, which are more likely to settle out. A more elaborate method of particulate removal is the use of electrostatic precipitators, which may remove up to 99 per cent of the particulates present. This involves the waste gases passing through electrically charged plates to which the particles are attracted. In this case smaller particles are just as likely to be removed as larger ones.

Sulphur dioxide is removed from exhaust emissions by a wide variety of techniques, collectively known as flue gas desulphurisation (FGD). Most of these involve the injection of an alkali into the gas stream, to which the sulphur dioxide will react, forming a solid waste product. The best-developed systems involve the use of limestone. This can either be used dry or wet. At present a number of smaller processes tend to use dry limestone, while larger plants have used wet techniques. Such FGD can reduce sulphur dioxide emissions by at least 90 per cent. Unless contaminated with other waste products (e.g. metals), the product of the cleaning is calcium sulphate (gypsum), which has some commercial value. The main disadvantage of limestone FGD is that very large quantities of limestone may be required, and concerns have been raised over the landscape impacts of its extraction. Other FGD techniques are also being developed. For large combustion plants in coastal areas, the use of the sea as a ready source of alkali is attractive. The wastewater, containing the sulphate (already of high concentration in the sea) can be discharged back to the sea. However, there are concerns over the potential impacts of other contaminants in the waste stream (e.g. metals).

Nitrogen oxides can be also be cleaned from exhaust gases. There are two principal means of achieving this – selective catalytic reduction (SCR) and selective non-catalytic reduction (SNCR). With SCR, a metal catalyst is used to reduce the nitrogen oxides to ammonia. It needs to operate between 200 and 400°C and can cope with a variety of other conditions (e.g. the presence of particulates). It can achieve an 80–90 per cent reduction in emissions. SNCR involves the injection of ammonia or urea into the flue gases to reduce chemically the nitrogen oxides. These reactions will only operate at high temperatures (900–1,100°C), so the technique is less versatile than SCR. SNCR can achieve a reduction in emissions of 70 per cent, although in practice lower figures are usually found.

Reducing ammonia emissions from agriculture

The most important source of ammonia emissions from agriculture is that from livestock waste. The ammonia may be emitted at any stage, from the production of the waste through to its final storage and use on the land. The pollution manager can address various stages in the process: nitrogen intake by animals, livestock housing management, storage of waste, and waste spreading (MAFF, 1992; Sommer and Hutchings, 1995). It is important to note that any controls on emissions from waste have to be made on a whole-farm basis. If emissions are contained within housing or waste, but not during spreading, then it will merely increase the potential for emissions later in the waste stream.

Reducing nitrogen intake by animals

Careful assessment of the protein requirements of livestock is an important means of reducing the amount of nitrogen excreted. By matching live-weight food requirements to nitrogen input, it is generally possible to reduce nitrogen intake by about 5 per cent. While most studies consider the role of processed feedstuffs, it is also important to consider grazing requirements. Thus high fertiliser application rates to grasslands can provide a surplus of protein available for grazing cattle and increase field emissions of ammonia.

It is also possible further to manipulate feedstuffs beyond simple management of protein content. For example, if pigs are fed a balanced diet of amino acids rather than crude protein, the amount of nitrogen in the waste is reduced. The slurry spread on fields results in a 40 per cent reduction in ammonia compounds in the soil, and a 60 per cent reduction in emissions to the air. The diet also has an added bonus of reducing the emissions of two potent greenhouse gases: methane and nitrous oxide. The cost to the farmer of the changed diet would add between £3.50 and £5.00 to the price of a pig by the time of slaughter. The adoption of the new diet would, therefore, depend upon general pig prices and on how readily the costs of dealing with ammonia pollution (to air and water) are passed on to the farmer.

Animal housing

Intensive animal housing can produce important point sources of ammonia emissions. A build-up of animal waste in moist conditions is ideal for pollution production. However, a wide range of techniques are now available for its control.

The slatted floors of animal units for cattle or pigs become covered in waste. Attempts to remove the waste through scraping have only a small effect on ammonia emissions. However, they may be flushed to remove the waste as it is produced and transfer it to contained storage. This may reduce ammonia emissions by about 50 per cent. However, it does increase the volume of the waste produced. Flushing efficiency can be improved by adding nitric acid to the water. This combines with the ammonia to form ammonium nitrate. The costs of doing this are partly offset by the reduction needed in further nitrogen fertiliser use after spreading of the waste.

For poultry units, droppings should be dried rapidly. Deep litter systems alone do not really reduce emissions. However, a reduction of up to one-third can be achieved by regular mixing of deep litter. For poultry kept in battery cages, droppings can be collected on conveyor-belt systems. If these droppings are rapidly removed to contained storage, a reduction in emissions of up to 60 per cent can be achieved.

Animal waste storage

It is generally not possible to use animal waste as it is produced. The waste is produced continually throughout the year, but spreading is limited, for example, to periods between cropping. Storage of waste is therefore required. Some waste storage does occur within animal housing, but much is separate. It is possible to achieve a reduction of between 70 and 90 per cent in ammonia emissions by simply covering the waste.

Waste spreading

The spreading of waste is the cause of about one-half of the ammonia emissions from agriculture. Spreading does also sometimes cause a public nuisance due to the odour produced in the spreading operation. However, contrary to public perception, the actual spraying of waste over land contributes to less than 1 per cent of the ammonia emitted. Most of the emissions take place within the first twenty-four hours, as the waste lies on the land surface.

The main control solution is to get the waste incorporated into the soil as quickly as possible. The simplest way to achieve this is to plough immediately after spreading. Given that most emissions occur soon after spreading, the ploughing must be undertaken quickly. A further solution is to use an injection technique whereby the liquid waste is passed directly into the soil in one operation. Such techniques are, of course, only applicable to fields where soil disturbance is acceptable.

The above techniques can result in a reduction in emission of up to 80 per cent. This also has direct benefits for the farmer, as the main reason for spreading waste is because of its value as a nitrogen fertiliser. Pollution reduction techniques can, therefore, reduce fertiliser costs to the farmer.

Managing pollution from motor vehicles

The problem of traffic pollution is a particularly difficult issue for the pollution manager. Traffic volume is increasing rapidly in almost every country in the world, and strategies to manage the resulting pollution are desperately needed. There are a number of types of management that can be adopted:

1   Changing the type of fuel used to one that produces less pollution.
2   Removing inefficient and grossly polluting vehicles.
3   Adopting measures to clean the exhaust gases.
4   Attempting to manage the pollution once it has been produced.
5   Adopting measures to manage the use of motor vehicles.

Some of these measures are best adopted at national or international levels, others are open to local management. However, in the long term, progress ought to be made on all of these options in order to achieve environmental targets in the most cost-effective manner. Each option will be considered in turn.

Managing fuel type

The best-known example of this has been the adoption of unleaded petrol. Lead is added to increase the octane level, and there are two ways to produce the same effect in unleaded fuel. The first is to improve the octane level of the petrol with a substitute for the lead, such as tertiary butyl ether (MTBE). This is, however, slightly more expensive than lead (by less than 1 per cent). The alternative is to change the refinery process so that a higher octane product is made at the outset. Catalytic reformation achieves this, but the higher-octane aromatic compounds do include substances such as benzene, which is a carcinogen.

Many countries have adopted strategies to reduce the lead in petrol. Its use is completely banned in the US, Canada, Austria, Sweden and some other countries. In the remainder of the EU, about half of the cars no longer use unleaded petrol (in 1994 UK use was about 58 per cent). In some cases unleaded fuel has been encouraged by the use of regulation. In the UK unleaded petrol was introduced in 1986. However, few drivers purchased it because it was marginally more expensive. However, the government changed the taxation regime to make unleaded petrol less expensive. This produced a rapid switch by most of those drivers whose cars could be converted. This was then supplemented by regulations (via the EU) requiring new cars to run on unleaded petrol and use catalytic converters.

However, many countries in the developing world still use large quantities of leaded petrol. The June 1996 Habitat II conference in Istanbul recommended that governments 'eliminate as soon as possible the use of lead in gasoline', but this is being resisted in some quarters. However, the World Bank is helping to overcome this with pilot projects in countries such as Thailand.

Fuel specifications can be adopted for other aspects of motor fuel use. Much of the effort of the EU has been to control the fuel quality of both petrol and diesel. Currently, petrol has a benzene limit of 5 per cent and this may be reduced to 1 per cent. There have also been progressive reductions in the sulphur content of diesel, which was 0.3 per cent in 1989, but in 1996 it was reduced to 0.05 per cent.

Consideration should also be given to proposals to change completely the type of fuel used. Biofuels are one option. Of these, the most extensively used is ethanol. *Biofuels* are produced from specially grown crops. In Brazil, for example, sugar cane is converted to ethanol as a substitute for petrol. Methanol from wood production is another alternative. Rape-seed oil can be converted to a methyl ester, which can substitute for diesel. Research into the use of biofuels is continuing. Their use is, however, still limited, except in Brazil where over 4 million cars now run on ethanol. Extensive use of biofuels would require large areas of land to be available for primary crop production. However, given the agricultural surpluses in many parts of the developed world, there may be opportunities for this. Biofuels not only have very low pollutant emissions, they could also be important in controlling greenhouse gas emissions, as the production and use of the fuel results in a low net carbon dioxide production.

Alternative fossil fuels can be used for motor vehicles, in the form of compressed gas, either from natural gas or liquid petroleum gas. Use is still very limited around the world, although it has been utilised on a small scale in Italy for around forty years, and it is here that nearly half of the world-wide number of gas-fuelled cars can be found. Gas still emits some nitrogen oxides, but carbon monoxide emissions are reduced by 90 per cent and there are no particulates. Where it is used it is often cheaper than petrol, but this tends to reflect the taxation environment rather than production costs.

Options are also open for *zero-emission vehicles*. In essence this means electric vehicles and the term 'zero emission' is only partially true. Thus while the vehicles themselves do not produce emissions, pollutants may be emitted by the power stations producing the original electrical power. However, some of the power stations may use renewable energy sources and it is, in any case, easier to control emissions from one large source than many small mobile sources.

The degree to which alternative fuels can be adopted relies heavily on consumer responses. For example, consumers need to be sure that supplies are available. This is a big problem for gas, electricity and biofuels in many countries where outlets may be limited. The take-up of unleaded petrol in the UK overcame this problem as petrol stations stopped supplying two-star petrol, and the tanks and pumps were immediately available for the new fuel. The use of fiscal incentives to encourage changes in fuel use (e.g. price differentials or car taxation rates based on engine type) is also important in overcoming consumer resistance and inertia.

Generally, countries rely on motor manufacturers and fuel producers to develop new solutions to pollution problems before they are incorporated into pollution management strategies. However, there is an argument for regulation as a 'technology forcing' mechanism. In other words, regulations stimulate the development of novel approaches and new technology. This is perhaps best exemplified by the state government of California. The problems of traffic pollution are particularly acute in a number of major cities, especially Los Angeles. The government has therefore adopted targets for zero-emission vehicles as a percentage of total car sales in the state until 2010. For 1998 this is set at only 2 per cent, but there are yearly targets up to 50 per cent in 2009 and 2010. Current technology cannot easily achieve this, but the incentive for manufacturers to develop and supply electric and other zero-emission vehicles to such a large market is intense.

Removing gross polluters

While many estimates can be given of emissions from different types of motor vehicle and fuel type, it is evident that older vehicles and those which are incorrectly maintained contribute a disproportionate amount to aerial pollution. For example, in a UK survey, Betts *et al.* (1992) found that 80 per cent of ordinary petrol cars needed tuning and 30 per cent of those which were fitted with catalysts. In 1994 the RAC reported that 44 per cent of the motor vehicle pollution in London was produced by 10 per cent of the vehicles. The use of older cars is also a reflection of prevailing economic conditions. For example, sales of used cars in the UK were 7.9 million in 1995, an increase of 14.5 per cent on 1994 and 43 per cent on 1992, and 2.4 million of these were over nine years old. This demand for older vehicles reflects the lack of consumer confidence in the UK economy in the mid-1990s and a resultant demand for cheaper motor transport.

In the US, active policies have been adopted to remove older cars from the vehicle fleet, using the accelerated vehicle retirement scheme. In the UK the aim is that gross polluters are discovered in the annual MOT test, and that engine adjustments have to be made before the vehicle can continue to be used. However, this test is only annual and many vehicles can quickly deteriorate or be re-adjusted after the test to achieve other objectives on performance. The UK has also adopted a 'smoky vehicles hotline', whereby the public can report diesel vehicles with visibly very high particulate emissions. In 1995–1996, for example, the vehicle inspectorate received 16,304 calls, with action being taken in more than half of the reported cases. Such action is most likely to involve a requirement that a vehicle pass an emissions test from an MOT centre. The UK's Vehicle Inspectorate has also undertaken a serious of high-profile emission 'blitzes' to identify gross polluters from the kerb-side through spot-checks. Under the 1995 Environment Act, local authorities in the UK are encouraged to set up a series of permanent kerb-side monitoring stations, especially where there are local air quality problems.

It might be considered that, gross polluters apart, the energy efficiency (and hence emission rate) of car use has increased considerably in recent years. However, this is not the case. While technological improvements have increased individual engine efficiency, the slight reduction caused by emission standard objectives (e.g. unleaded petrol and catalytic converters) and the greater demand for higher-performance cars has partly offset this. However, if measured as the energy used per mile travelled per passenger, there has been no improvement in the UK in the last 25 years. In fact there has been a slight decline (DoE, 1996). Much of this has been due to a large reduction in the number of people travelling in each vehicle and the fact that in many cities a considerable amount of fuel is used while cars are trapped in traffic congestion.

Cleaning exhaust gases

Due to difficulties in producting vehicles that produce little or no emissions of toxic pollutants, much of the recent effort on pollutant management has focused on 'end-of-pipe' technology, i.e. cleaning the pollutants from the exhaust gases.

Catalytic converters have been fitted to cars in Japan and parts of the US for over twenty years. There are two types available. The first is a simple porous ceramic structure coated in a catalyst (usually platinum), through which the exhaust gases pass. The catalyst converts carbon monoxide and hydrocarbons to carbon dioxide and water vapour, but does not affect nitrogen oxides.

The three-way catalyst combines the use of different catalytic metals. Platinum is still used to oxidise carbon monoxide and hydrocarbons, and rhodium is used to reduce the nitrogen oxides to nitrogen, carbon dioxide and water. This efficiency of this type of catalyst is particularly sensitive to the air/fuel ratio in the engine (the optimum is 14.7:1), and most cars are now fitted with a lambda sensor in the exhaust to monitor emissions and feed back to the fuel injection systems. All current regulations in the EU now require manufacturers to fit three-way catalysts.

Both types of converter require efficient functioning of the sensitive metal catalyst. Thus unleaded fuel must be used, as lead rapidly interferes with the catalyst. The catalysts also operate most efficiently at high temperatures (around 300–400°C). This may take several minutes to achieve. Recent studies on average journey times show that cars are increasingly being used for very short journeys, especially in towns and cities. Thus, even if all vehicles were fitted with catalytic converters, they would still emit significant quantities of toxic pollutants into the urban environment.

Catalysts do not deal with the problem of particulate emissions. Most transport particulates are produced from diesel vehicles. It is possible to fit a simple particulate trap in the exhaust. However, these traps quite rapidly become clogged with particles and back pressures can build-up in the engine. Currently, research is being undertaken to examine ways to regenerate these traps. One option is to burn off the particles. However, this requires high temperatures or the use of catalysts, either in the exhaust stream or as a fuel additive.

In the UK, catalysts became compulsory for new cars at the beginning of 1993. The UK government has estimated that catalysts will lead to a reduction in nitrogen oxide emissions of 35 per cent by the year 2000 and 52 per cent by 2010, from the 1993 baseline. An early test of the effect of catalysts may be found in results from annual nitrogen dioxide monitoring. So far only results from 1993 and 1994 are available (NETCEN, 1996). In 1994 it was estimated that nitrogen oxide emissions should have declined by 1.1 million tonnes due to the effect of catalysts. However, surveys of ambient air quality showed an increase in nitrogen dioxide levels in urban areas for kerb-side (5 per cent increase), intermediate (3 per cent) and background (1.4 per cent) situations. There may be a number of reasons why the control measures have not translated quickly into air quality improvements, but it suggests that there should be caution in predicting how quickly air quality might improve with the adoption of these measures. It would also be possible to encourage retro-fitting of catalytic converters by using tax incentives, and this has been adopted in Germany this year (1997).

The introduction of catalytic converters and other pollution control measures does add to the cost of motoring. For example, the European Auto-Oil Programme estimated that it would cost about £2.75 billion across the EU to achieve a reduction of 40 per cent in nitrogen oxides and 30 per cent in particulates. This is equivalent to £180 for each motor car and £1,330 for each heavy-goods vehicle. Passing such costs on to the consumer is, of course, consistent with the polluter pays principle (see Chapter 1).

Direct management of pollution

It is even possible to attempt to remove the traffic pollution once it is produced, though in reality this ought to be the last management option considered. One city,

Paris, has examined this issue. Like many major cities, Paris has severe air pollution problems. During the winter of 1995/6, the transport workers' strike resulted in greatly increased traffic levels and record concentrations of air pollutants. It is also estimated that about 350 people die every year from coronary and respiratory diseases related to air pollution. The Ministry of the Environment has examined different ways of dealing with the air pollution problems in Paris. A key feature of many urban air quality problems is the presence of extreme episodes resulting from high pollutant emissions and unfavourable climatic conditions. The Ministry aims to overcome these by installing 70,000 electric fans on buildings throughout the city. These will all be coordinated by one centre that will monitor air pollution and climatic conditions. They will switch on when pollutants are at an unacceptable level and create a breeze of about 5 km/h to 'blow' the pollutants away. The cost of the installation of the fans is estimated at £52 million. Their success will partly depend on whether their noise is considered acceptable by those living and working in the buildings on which they are placed. It is also important to monitor how cost effective these measures are in comparison with other pollution reduction strategies in other European cities (e.g. traffic management). Wider monitoring of impacts (e.g. of subsequent rural pollution levels) is also important.

Another alternative has been proposed to treat the air in Los Angeles. This involves the construction of chimneys 150 m wide, where sea-water would be pumped to the top and be sprayed out into a fine mist. This would evaporate, cooling the air and dissolving the pollutants. A cool breeze would blow from the bottom of the chimneys, displacing the city's polluted air. Current estimates are that about 100 chimneys, costing US$1 billion, would be needed to clean the air of Los Angeles.

Traffic management

One alternative to prevention and management of traffic emissions is to reduce the need for or the use of motor vehicles. Two issues can be considered: vehicle speed and vehicle use.

The quantity of pollutants emitted per kilometre driven varies with the speed of the vehicle. Figure 3.2 shows how emissions vary for non-catalytic petrol cars. For example, when cars travel much faster than 70 km/h, nitrogen oxide emissions begin to increase. Thus the efficient enforcement of speed limits on motorways would help to reduce pollutant emissions. There are even technical solutions to this, whereby speed regulators can be fitted to cars (as they are to some heavy-goods vehicles and coaches for safety reasons) to prevent speed limits being exceeded. However, political lobbying has prevented the use of these, even though their lack of use encourages illegal activity.

Figure 3.2 also demonstrates the problems of emissions at very low speeds. Thus as congestion in towns and cities increases and traffic slows down, emissions are increasing much faster than the actual growth in vehicle numbers. This requires a management response to reduce vehicle use and to keep moving those vehicles that *are* used.

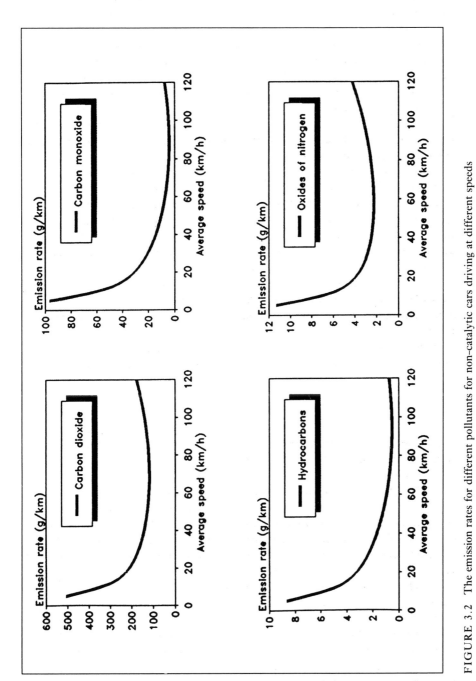

FIGURE 3.2 The emission rates for different pollutants for non-catalytic cars driving at different speeds
Source: DoT (1994).
Reproduced with permission of the Controller of Her Majesty's Stationery Office.

To keep vehicles moving requires the creation of clear through-routes and precise management of road junctions such as traffic lights and roundabouts. Creation of clearways (no parking) and their enforcement aids traffic movement. Detailed models are now available to assess traffic flow rates along roads and at intersections, and to determine the optimum management response for peak usage. These models can also be linked to air pollution models to determine whether vehicle flows are liable to produce peak pollutant levels likely to pose a health risk in adverse meteorological conditions. A management response (e.g. prevention of vehicle use) could then be implemented. Under the 1997 National Air Quality Strategy, as a trial, the UK government is to give five local authorities the powers to assess air quality problems, test polluting vehicles and close heavily polluted roads. If the scheme works well, this will be extended to the whole country.

Various methods can be used to reduce vehicle use. Land-use planning decisions can be made to discourage private car use. The creation of pedestrian zones, specified cycle and bus lanes/roads and adequate provision of public transport for key services all play a part. Disincentives for car use may involve road pricing schemes. 'Smart' cards can be used, which have microchips within the cars and which are registered by sensors along the roads to record use. Drivers then pay for their road use. Such a system operates in Oslo and is being investigated for London. General fiscal regimes, e.g. heavy taxation on vehicles, can discourage ownership. For example, the current UK tax system heavily favours the use of company cars, but for political reasons neither the current government nor the main opposition have proposals to change this. However, more success is likely to be achieved in discouraging car use rather than ownership itself (which tends to discriminate against low-income groups). Some cities, e.g. Mexico City, have adopted schemes whereby cars are only allowed to enter the city on certain days of the week according to the last digit of their number plates. This encourages car sharing between those with cars having different codes. However, the scheme has been less successful than predicted as some drivers have bought older, more polluting, cars to use during days when they cannot drive their main vehicle. The character of vehicle fleets in Latin America is very different from North America or Europe. In São Paulo in Brazil, for example, the current fleet size of 5 million is growing at the rate of 250,000 per year, but few old vehicles are being replaced, so that older, polluting vehicles are growing in number. In August 1996 São Paulo introduced a pollution control scheme similar to that in Mexico City, backed by heavy fines. About 95 per cent of motorists obeyed the controls, but pollution only fell by 14 per cent, as the most polluting vehicles (lorries and buses) were exempt from control.

All of these methods are open to those managing traffic pollution. However, the management of traffic numbers is difficult because it goes against trends in popular culture. People now consider car use to be almost a 'right' and demand an ease of mobility that was unobtainable a few years previously. Thus the most important task is one of public education and lifestyle change and, of all the means available for attempting to reduce traffic pollution, this is most definitely the hardest, but the most fundamental.

Environmental assessment

For many pollution managers the first examination of a potential air pollution problem will come in the preparation of an environment statement for industrial processes or roads that may emit air pollutants. Below is a checklist of issues that should be considered when preparing an environmental statement. Not all are relevant to every source, and common sense is required in the amount of detail needed.

1   Pollution emission data. Presentation of different aspects of a pollutant emission may be necessary. First the concentration (mg/m^3) and emission rate (g/s) in flue gases or from motor vehicles should be given. This is probably all that is necessary for small processes. However, if these rates are high, modelling of ground-level concentrations ($\mu g/m^3$) will be necessary (see 2). Finally, if total pollution emissions are large, a measurement of the total mass emitted (t/yr) is necessary in order to assess the contribution of the process to regional pollutant deposition.

2   Modelling of emissions to estimate effects on ground-level air concentrations should be done under a range of weather conditions. Health concerns largely centre on short-term peaks, whereas impacts on buildings and the natural environment are better understood for long-term doses of pollutants. It is, therefore, necessary to know annual mean concentrations and peak concentrations, the latter being expressed in the time intervals that have been set as air quality standards for the protection of human health. If operation of the process is unlikely to be uniform throughout the year, it is necessary to know the effect on ground-level concentration during those periods of operation. Similarly, peak concentrations from roads should reflect periods of greatest use.

3   The historic and possible future trends in ground-level air concentrations should be determined. Buildings and the natural environment will have been affected by the past pollution history of an area. Thus a small increase in pollutant concentration may have no impact in an area that has shown a significant recent decline in background levels, but in a different location it could cause a critical level to be exceeded if background levels have been rising. Assessing future trends will be necessary for processes in areas where other major new developments are under construction or consideration.

4   For pollutants emitted in significant quantities it may be necessary to estimate deposition rates and compare these with background deposition. This is particularly so for nitrogen oxides and ammonia, which may be of low enough concentration as to not cause direct toxic effects through the gases entering leaves, but could cause a significant increase in the nitrogen status of a sensitive community. Similarly, low levels of sulphur dioxide may cause no health problems, but may result in deposition that affects buildings.

5   The presence of other pollutants must be considered. Health impacts may be increased by pollutant mixtures, and natural systems may respond to multiple pollutant stresses. Even if a process is only emitting one significant air pollutant, it is only possible to assess its impact if background levels of other pollutants are

known. Thus sulphur dioxide, nitrogen dioxide and ozone may all have direct synergistic toxic effects. Sulphur dioxide and ammonia will also deposit at a higher rate if both gases are present together.

6   It is important to describe the environmental receptors in the vicinity of the process. This should include an indication of any local populations and sensitive sites (e.g. schools and hospitals) and important components of the natural environment (e.g. nature reserves and rare species).

7   For processes that emit very large amounts of pollutants, it is necessary to assess regional or even national impacts on the contribution to total acid deposition. This should be compared with maps of acid-sensitive areas and critical loads maps.

8   If there is a possibility that some impact on environmental receptors may occur, the effects of alternatives to the proposed process should be given. This may be as simple as to indicate the effect of increased chimney height, the effects of different fuel options, options for removal of pollutants from flue gases, or alternative designs/routes for roads.

9   Finally, after such analyses, applications for processes that may present a potential problem should indicate ways that environmental monitoring (both pollutant and biological) will be undertaken to assess whether impacts do occur.

Chapter summary

◆   This chapter examines the history of legislation controlling air pollution, which goes back many centuries. Current UK legislation is dominated by the 1990 Environmental Protection Act implementing Integrated Pollution Control (IPC) through the concept of best available techniques not entailing excessive cost (BATNEEC). The operation of IPC to all media is described, as well as techniques that may assess how different impacts in the different media may be compared.

◆   The role of air quality standards in air pollution management is described, with the limitations of current standards being stressed, i.e. they tend to be set only to protect human health and not the wider environment. Some are operated through EU legislation and others through the 1997 UK national air quality strategy.

◆   The need for international controls is stressed, as air pollutants do not respect frontiers. The EU has produced many directives to set emission limits, monitor pollutants, standards, etc. In 1996, an important development was the agreement on a Directive on Integrated Pollution Prevention and Control (IPPC), which extends pollution control beyond IPC to new processes and sets the regulation in a wider context. Even the EU is not a sufficiently wide framework for some controls, e.g. acid rain, and the role of the UN in this is described.

◆   An important aspect of managing air pollution is the need to monitor its distribution and trends over time. A range of methods are described, varying in their accuracy and cost. It is also important to model dispersion, and the nature of models is described, together with an examination of a simple Gaussian dispersion equation.

◆ Three areas of pollution management are examined in more detail:

1 *Industry*. Different methods are described covering: changing fuel type, changing process operation and cleaning flue gases.
2 *Agricultural ammonia*. Different methods are described, covering: nitrogen uptake by animals, housing conditions, waste storage and spreading.
3 *Traffic pollution*. Different methods are described covering: changing fuel type, removing gross polluters, cleaning exhaust gases, directly managing air quality, and traffic management.

Recommended reading

The reading recommended in Chapter 2 also covers a range of regulatory and management issues.

Allott, K. (1994). *Integrated Pollution Control: The First Three Years*. ENDS, London. This is a detailed examination of the operation of the 1990 Environmental Protection Act in Britain. It examines the way that the legislation has been implemented, the information presented, enforcement actions and the role of other groups, such as the public.

DoE (1997). *The UK National Air Quality Strategy*. HMSO, London. This sets out in great detail the air quality problems facing human health in the UK. It examines the nature and role of standards and sets out a management strategy to deal with these in the context of existing and forthcoming UK and EU legislation. This document contains much useful background information.

EAC (1996). *Released Substances and their Dispersion in the Environment*. HMSO, London. This is a joint industry/regulator publication, which examines the information required in assessing the impacts of pollution from industrial processes. It reviews the nature of pollution dispersion and the models that can be used to predict it. It is presented in a clear manner, to aid individual pollution managers.

The freshwater environment

◆ Introduction   **104**
◆ Threats to freshwater   **104**
◆ United Kingdom and European Union
  legislative background   **105**
◆ A strategy for water quality in England and
  Wales   **109**
◆ Sources of pollution in freshwater   **112**
◆ Groundwater   **149**
◆ Chapter summary   **151**
◆ Recommended reading   **152**

Introduction

The water resources of the Earth, known as the *hydrosphere*, includes not only fresh-waters and the oceans, but also the ice caps and glaciers. Only 1 per cent of the water on the surface of the Earth is freshwater. Freshwater is vital to almost all terrestrial life (and, of course, freshwater life), including mankind, and thus the protection of this limited resource from pollution is vital. Overall, across the Earth, the majority of freshwaters are not so polluted as to prevent their use by humans or wildlife. However, significant pollution threats do occur to the freshwaters in particular areas of the world. The issue of pollution also has to be considered alongside other threats to the resource.

Threats to freshwater

The freshwater resource on which mankind and the wider biosphere depends, is under threat from a variety of pressures. Many of these are driven by man, although a few are natural (RCEP, 1992). This volume is only concerned with pollutant impacts within the freshwater resource. However, it is important to note that the threat of water loss (e.g. abstraction from groundwaters or rivers), leading to low flows and impacts of recreation on wildife and habitats, are also serious concerns. These threats may interact with pollution (e.g. discharges to low-flow rivers will have less dilution and so result in higher *in situ* concentrations of pollutants). The pollution manager must, therefore, adopt an integrated approach to the management of fresh-water pollution and, therefore, work with all those concerned with the management of freshwater catchments.

This chapter will initially examine the legislative and regulatory framework in which the pollution in freshwaters is managed. For surface waters, it will then examine the individual pollutant types in turn, including pathogens, metals, organic pollution, suspended solids, nitrates, eutrophication, acidification, thermal pollution and pesticides. In each case, sources will be examined, with assessments of impacts and options for pollution managers to prevent pollution or manage its effects. Case studies will be given to illustrate many of the key issues. Finally, pollution in ground-waters will be examined briefly.

United Kingdom and European Union legislative background

Early controls in the UK

Serious consideration was given to the issue of river pollution in the UK in the mid-nineteenth century, when the River Pollution Commission published its recommendations in 1868. These included the creation of regional boards to oversee pollution control, with inspections and standards for effluent discharges. However, the subsequent legislation, the River Pollution Prevention Act 1876, failed to act on these recommendations, and it was not until the mid-twentieth century that such measures were adopted. Thus, while some legislation (e.g. the Public Health Act 1937) enabled spot-checks by local authorities in industrial premises, nearly one hundred years after the River Pollution Commission, many rivers were still no less polluted. Thus while the Thames, for example, is now able to support salmon, other lowland rivers still suffer from sewage and agricultural pollution, or from discharges from recently abandoned mines.

Modern UK legislation

A series of acts aimed at preventing river pollution were passed between 1951 and 1963. These acts introduced a consent-based system of pollution control, with procedures for sampling and other inspection. The regulating authorities could require conditions on discharges, e.g. volume, site of outfall, etc. However, many consents were unconditional, and environmental bodies criticised these as being a right to pollute. The legislation was again revived in the 1974 Control of Pollution Act (COPA), when controls began to tighten and river water quality in many areas could be seen to begin to show a significant improvement.

European Union directives

Much modern UK regulation derives from the European Union. There have been a large number of directives within the EU designed to improve water quality. As with air quality, these directives deal with quality issues from a variety of different angles. For example, there are regulations concerning the discharge of particular substances: there are objectives set for water quality itself; and there is the beginning of regulation of particular industrial sectors. However, again as with air pollution, the European philosophy differs from UK practice, with Europe laying greater emphasis on meeting particular limits on discharge, while the UK focuses on environmental objectives – with a more flexible approach to how these are met. Important EU directives include:

Controlling discharge of dangerous substances

The 'Framework' Directive (76/464/EEC), and its subsequent directives, seek to control discharges of dangerous substances to surface waters and the directive on Protection of Groundwater (80/68/EEC) does the same for groundwater.

The Framework Directive provides a list of substances that are considered to be so toxic that it is considered necessary to prevent pollution by them. This list is termed the 'Black List'. Substances in this list include DDT, mercury and cadmium. The control on discharges is achieved by defining limits on the concentrations of these pollutants in discharge outlets and also setting limits for water quality. Many of these substances are incorporated into the UK's Integrated Pollution Control regulation as prescribed substances, and in this case are known as the 'Red List'.

The Framework Directive also provides a second list of substances, known as the 'Grey List', which are less toxic than the 'Black List' substances. The Directive requires that member states draw up their own strategies to reduce pollution by these substances, which include some metals such as chromium and nickel, and other pollutants such as ammonium and cyanide. In the UK, the government response has been to produce a circular outlining environmental quality standards for all 'Grey List' substances. The accompanying box outlines the substances included in the two lists.

BOX 4.1 The black and grey lists of substances from the European Union Dangerous Substances Directive

Black list Mercury, cadmium, hexachlorocyclohexane, DDT, pentachlorophenol, carbon tetrachloride, aldrin, dieldrin, endrin, isodrin, hexachlorobenzene, chloroform, trichloroethylene, tetrachloroethylene, trichlorobenzene.

Grey list 1,2-Dichloroethane, lead, chromium, zinc, copper, nickel, arsenic, boron, iron, pH, vanadium, tributyltin, triphenyltin, PCSDs, cyfluthrin, sulcofuron, flucofuron, permethrin.

The Groundwater Directive builds on the Framework Directive, requiring that 'Black List' substances are not discharged to groundwater and that the potential impacts of 'Grey List' substances on groundwater should be studied before their discharge is permitted.

Water quality objectives

A number of directives regulate the quality of the aquatic environment. Two of these are aimed at regulating water quality *in situ*. However, these are not comprehensive in their coverage of the environment, being aimed specifically at drinking water and bathing water. The third directive acts to improve water quality by setting standards for water treatment. It is necessary that the problem is approached from both sides, and this is what the EU is doing. However, it is not, as yet, setting out an over-all strategy for defining the comprehensive pollution problems associated with the freshwater environment and therefore the suite of measures needed to deal with them. These directives are as follows.

The Surface Water Directive (75/440/EEC) classifies surface waters according to the treatment necessary for use in drinking water:

1   A1 – requiring only basic filtration and disinfection;
2   A2 – requiring standard physical and chemical treatment;
3   A3 – requiring extensive physical and chemical treatment.

The companion to the Surface Water Directive is the Drinking Water Directive (80/778/EEC), which defines the quality of water necessary for drinking. For a wide range of substances, three values may be given. These are Guide Levels, Maximum Admissible Concentration and Minimum Required Concentration. Of these three, the first is not legally binding, but acts to direct the work of pollution managers. The levels are desirable goals. Failure to meet them is, of course, likely to lead to public disquiet. The latter two sets of values are binding. These may apply to toxic substances such as lead and health hazards such as bacterial contamination. Some Maximum Admissible Concentrations are outlined in Table 4.1. They also apply to factors affecting taste and to substances that cause problems in high concentrations, e.g. nitrates. Minimum Required Concentrations apply to treatments for water softening.

The Bathing Water Directive (76/160/EEC) covers both freshwater and marine areas. However, as most concern has been over implementation on marine

TABLE 4.1 Maximum Admissible Concentrations for drinking water for a range of selected contaminants as defined in the EC Drinking Water Directive

Contaminant	Maximum Admissible Concentration
Acidity	pH range 5.5–9.5
Turbidity	20 units
Colour	4 units
Aluminium	200 µg/l
Iron	200 µg/l
Lead	50 µg/l
Manganese	50 µg/l
Nitrate	11.5 µg/l
Pesticides (total)	0.5 µg/l
E. coli	Not detectable in 100 ml of water

Source: EC 80/778/EEC

beaches, it will be discussed in Chapter 5. However, it is important to note that in achieving adequate marine water quality, the greatest effort may be needed when attempting to reduce the pollution entering rivers that ultimately discharge into the sea.

The Municipal Waste Water Treatment Directive (91/271/EEC) sets minimum standards for the treatment of municipal wastewater. This includes discharges of industrial wastewater, which may be considered similar to municipal wastewater. The control is based on the quality of the waters that will receive the discharge, and a particular aim is to prevent the eutrophication of the receiving waters when nitrates and phosphates are discharged.

Nitrate pollution is also regulated under a separate directive (91/676/EEC), which seeks to reduce nitrate pollution problems from agricultural sources. Large quantities of nitrates are applied as fertilisers in modern agriculture and these ultimately enter surface waters and groundwaters. Drinking water requirements are for a concentration of only 50 mg/l of nitrate. If catchments exceed this concentration then they are considered as 'nitrate vulnerable zones'. A series of measures are then open to member states to improve the water quality. These include reductions in fertiliser application rates, physical measures to prevent run-off and reduction in use close to watercourses.

Future directions in the European Union

It is difficult to predict where water quality issues will proceed within the EU. Two general areas of policy/regulation are important to set the future in context. The first is the forthcoming Directive on Integrated Pollution Prevention and Control. This is described in Chapter 3, but it is important to note that, for many major sources of pollution affecting freshwaters, a tighter regulatory regime is in prospect. The second area of regulation is that policy has to be set within the framework provided by the EU's Fifth Environmental Action Plan – 'Towards Sustainability'. The proposals for legislative change that are inconsistent with moves towards achieving sustainable development are likely to be fiercely resisted (especially by three of the newer member states who are more environmentally 'aware' – Austria, Finland and Sweden).

Some member states (e.g. France or Britain) are concerned about some of the implications of existing directives and may be looking to the agreement on subsidiarity (i.e. devolution of decision-making to member states) within the Maastricht Treaty to redress this. However, many commentators consider that such developments are unlikely. It is probable that both the drinking and bathing water directives will be revised, but this is most likely to take account of changing scientific knowledge, which may move standards up or down, but not relax the protection for human health. There may also be additional moves to examine sources of *diffuse pollution*, as most point sources are covered in some measure already. Protection of surface waters from acidification will also be included within the current development of an acidification strategy for the EU.

Recent UK developments

The 1989 Water Act privatised the water industry and enabled a complete separation from the water companies (responsible for water supply and sewage treatment) from the regulator (the National Rivers Authority) who could regulate their activities. The 1995 Environment Act subsequently incorporated the NRA into the Environment Agency. The 1989 Act provided clearer frameworks for setting consents. This included charging schemes based on the nature of the substances being released, and a strategy for environmental objectives.

Until now, the discharge limits on industrial processes have been set using environmental quality standards (EQSs). The aim is to set an acceptable concentration of a pollutant in a river (the EQS) and estimate the potential concentration resulting from a discharge. A breach of an EQS would be unacceptable. This was undertaken in the context of an overall assessment of pollution to land, air and water (Integrated Pollution Control), and in 1996 was expanded to Integrated Pollution Prevention and Control, following the 1996 IPPC EU Directive. Both of these systems are described in detail in Chapter 3. The system of assessing discharges against EQSs is also likely to change to one based on a direct assessment of the toxicity of the effluent being released, irrespective of what it contains. Direct toxicity assessment (DTA) is equally applicable to freshwater and marine discharges and is considered in detail in Chapter 5.

A strategy for water quality in England and Wales

Although regulation of freshwater pollution is now incorporated within the Environment Agency in England and Wales, it is too soon after its formation for significant new strategic developments to have taken place. The predecessor body for water protection, the National Rivers Authority, formulated a strategy (NRA, 1993) and it is therefore appropriate to consider this here. It has two principal aims and five subsidiary aims needed to achieve them. These are as follows.

Principal aims

1   To achieve a continuing overall improvement in the quality of rivers, estuaries and coastal waters through the control of pollution.
2   To ensure that dischargers pay the costs of the consequences of their discharges.

Subsidiary aims

To achieve these aims the NRA seeks to:

1   Maintain waters that are already of high quality.

2   Improve waters of poor quality.

3   Ensure all waters are of an appropriate quality for their agreed uses.

4   Prosecute polluters and recover the costs of restoration from them.

5   Devise charging regimes that allocate the costs of maintaining and improving water quality fairly and provide incentive to reduce pollution.

River management in England and Wales

The Environment Agency in England and Wales has adopted a fully integrated approach to river catchment management, including that of water quality. The tool for this management was, until recently, the preparation of Catchment Management Plans, which were being, or had been, prepared for the 163 principal catchments in England and Wales. However, the new Environment Agency has replaced these with Local Environment Agency Plans (LEAPs), which include plans for waste regulation and air pollution. However, the principles of integrated protection of the water environment developed in Catchment Management Plans will still be found in the LEAPs. The principal aim of the plans is to examine the interactions (physical, biological and human) between the aquatic environment (groundwaters, running and standing waters and the estuarine and marine environment) and the land uses that take place in the catchment. This approach identifies the main problems facing the aquatic environment, and the main causes of the problems. The integrated approach allows more cost-effective solutions to be identified and reduces the likelihood that solving one problem creates another. Problems may arise with a lack of information relating to some sectors (e.g. pollution, fisheries, etc.) or a dominance of interest by other sectors. However, these practical problems are outweighed by the benefits that can be achieved by bringing disparate interests together.

The main issues that were contained in a Catchment Management Plan are summarised in Table 4.2. Many of these directly relate to water quality (e.g. discharges and agricultural land use) and others indirectly relate to it (e.g. a river with a low flow will have more concentrated pollutant levels as its capacity to dilute received discharges is reduced).

Rivers in Britain are classified according to their water quality. The rivers and canals of Britain are surveyed every five years by the Environment Agency and lengths of each are allocated to quality bands in a classification system. The rivers and canals can be classed as:

1   good (1A and 1B); these are suitable for extraction for drinking water, maintain high-class fisheries and have a high amenity value;

2   fair (2); these have similar purposes to those of class 1, but with lower general water quality;

3   poor (3); these may not support fish life;

4   bad (4); these are grossly polluted and may cause a nuisance.

In a LEAP, each stretch of river is classified as above. Different use objectives can be

TABLE 4.2   Main issues addressed in a Catchment Management Plan

Sector or major land use	Issues to be addressed for catchment management
Water transfers	River flows Chemical composition
Groundwater abstraction	Groundwater levels Pollution of sources River flows
Abstraction for agriculture, industry and domestic supply	River flows Water quality
Agriculture and forestry uses	Run-off Pollution Siltation
Housing, industry and urban development	Demands for potable (drinking) and non-potable water Discharge of surface water Volume and nature of effluents
Wastewater treatment	Volume, nature and location of effluent Discharges Dilution of effluent
Land drainage and flood defence	Groundwater levels Floodplain and coastal marshlands Management Flood flows
Mineral extraction and waste disposal	Pollution
Fish farming	Local flows Pollution
Fisheries and angling	Water quality River flows Channel character
Boating and commercial navigation	Channel shape and depth Locks and moorings Facilities Pollution
Bankside recreation	Public access needs

Source: After English Nature (1995)

111

set for different parts of the catchment. These can either reflect the current state of water quality (e.g. polluted water may be less suitable for recreation), or reflect a desired aim, which management practices can be set in place to achieve (e.g. improve effluent treatment at a sewage treatment works or change farming practices). Under the Water Resources Act (1991) a system of 'Statutory water quality objectives' (SWQOs) will be introduced to meet these needs. The use classifications will be based on the following criteria:

1   Overall river ecosystem.
2   Abstraction needs for drinking water.
3   Abstraction needs for industry and agriculture.
4   Recreational needs.
5   Special ecosystem.

The real challenge for SWQOs in LEAPs is whether they will merely reflect the status quo, i.e. whether they will be descriptive of the current water quality along a river and thus be management tools to maintain this, or whether they will set serious objectives beyond the current river status in order to achieve long-term environmental improvements.

Sources of pollution in freshwater

It is simple to list the different sources of pollution to freshwaters. However, this can present a misleading picture of the relative importance of these different sources. In Britain the Environment Agency undertakes an annual survey of all pollution incidents to freshwater in England and Wales and an examination of the data reveals where the primary pollution sources are. Table 4.3 provides summary data for a sample year – 1993.

This survey demonstrates the enormous range of sources of pollution in fresh-water. While many of these will produce similar types of pollutants (as categorised later in this chapter), the relative importance, size, timing and social and political context in which they operate vary enormously. The pollution manager therefore has the difficult task of tackling these diverse pollution sources.

Pollutants in freshwater

Pathogens in freshwater

The discharge of untreated or poorly treated sewage from both human and animal sources can lead to freshwaters containing a wide variety of human *pathogens*, and these can pose serious problems to human health (Abel, 1989). These organisms can range from viral diseases such as hepatitis, bacterial infections to protozoans such as *Entamoeba*, and can also include parasitic worms such as tapeworms and *Schistosoma*.

TABLE 4.3   Summary of the sources of pollution incidents recorded in England and Wales during 1993

Pollutant source	Number of incidents	Pollutant source	Number of incidents
Agriculture		*Sewage and water related*	
Dairy cattle	1,007	Combined sewer overflows	1,422
Beef cattle	917	Surface water outfalls	1,255
Other	568	Sewage treatment works	1,348
Pigs	194	Foul sewer	1,001
Poultry	39	Pumping stations	481
Arable	51	Other	255
Fish farms	31	Cess pits	243
Horticulture	20	Unclassified	214
Stables	17	Water treatment works	67
Sheep	10	Water distribution	33
Mixed	6	Sewer dykes	24
Small-holding	2	Storm tanks	19
Forestry	1	Rising mains	11
Industry		*Transport*	
Oil industry	1,288	Road	1,092
Chemical industry	601	Ships	236
Construction	353	Unclassified	105
Landfill/waste disposal	328	Pipelines	36
Mining	241	Rail	21
Food industry	126	Airports	8
Demolition	85		
Engineering	63	*Other*	
Paper industry	43	Domestic/residential	748
Fuel stations	18	Crown exempt	24
Power generation	11	Restaurant/hotel	17
Textile industry	4	Scrapyards	9
Metal industry	3	Hospitals	9
		Contaminated land	8
		Schools	7

Source: NRA (1994)

Table 4.4 provides some examples of the types of disease transmitted in infected waters.

The potential for infection increases with the increasing concentration of human populations. In many parts of the developing world where sewage treatment is inadequate, the rapidly developing urban areas are particularly at risk. As with many diseases, it is the young, old and sick who are most at risk from these infections.

TABLE 4.4 Examples of the range of diseases transmitted in infected waters, according to type of organism and symptoms caused

Infective agent	Symptoms
Viruses	
Echo	Meningitis, respiratory disease
Hepatitis A	Hepatitis
Polio	Paralysis, meningitis
Rotavirus	Diarrhoea
Bacteria	
Cryptosporidium	Diarrhoea
Escherichia coli	Diarrhoea
Salmonella typhi	Typhoid
Salmonella spp.	Gastro-enteritis
Shigella spp.	Bacterial dysentery
Vibrio cholera	Cholera
Protozoan parasites	
Entamoeba histolytica	Amoebic dysentery
Giarda lamblia	Severe gastro-intestinal infection
Invertebrate parasites	
Blood flukes (*Schistosoma* spp.)	Serious debilitating illness
Roundworms (e.g. *Trichuris trichiura*)	Dysentery
Tapeworms (e.g. *Taenia saginata*)	Debilitation

Note: The selection is world-wide. There are very few temperate instances of human infection from drinking water for polio or hepatitis A and some invertebrate parasites

There are a number of methods for reducing the spread of disease in drinking water. The most obvious is that, if sewage cannot be treated, it should be discharged below the source of water for drinking and recreational use. The diversion of London's sewers through the city to discharge in the lower part of the Thames in the late nineteenth century played a great part in improving the health of the population in the city. However, problems in the UK due to contaminated water are not just a thing of the past. In 1989 500 people in Swindon suffered chronic diarrhoea from contamination with *Cryptosporidium*, as did 300 people in Devon in 1995. Thus there is a need for continued treatment and vigilance.

However, proper management of the problem requires adequate secondary treatment of sewage. The main course of treatment is to disinfect the water with chlorine or, less commonly, with ozone. Both methods work more efficiently once suspended solids are reduced to a minimum and at optimum pH. The degree of effect is increased with increasing residence time of the treated water. The length of treatment required will depend on the nature of possible infections. For example, coliform bacteria are more easily killed than viruses, and protozoan cysts are the most persistent. For other treatment methods, the stage at which disinfection takes place needs to be assessed. For example, treatment with chlorine cannot precede further secondary biological treatment for ammonium (see p. 127) as the beneficial bacteria will be killed.

The detection of biological contaminants (for example, to assess the need for or success of water treatment) can be a difficult process, especially if they are in low concentrations. One development that may result in a quicker and easier way to detect a range of organisms is being tested at the University of Wales in Bangor. This involves small silicon chips which contain a series of 'conveyor belts' made up of a series of electrodes. Each organism creates its own electrical field, and the electrodes can be used to trap organisms with particular fields. Identification can be further speeded up using small beads labelled with antibodies specific to individual species. If this is successful, water companies and other users will be able to detect individual organisms and receive results in only minutes, instead of days.

BOX 4.2 Biological monitoring

Surface waters can be monitored in various ways. Most obviously they can be assessed directly for the presence of pollutants of concern, e.g. by monitoring for pathogens or phosphate. If these pollutants are liable to change rapidly over time (e.g. daily variations or maybe as single events), then frequent (and thus expensive) monitoring is required. An alternative approach for some pollutants is that of biological monitoring.

Biological monitoring has been developed by examining the detailed biological interest of different rivers or lakes with known water chemistry (Hellawell, 1986). For example, individual species of mayfly or caddis-fly larvae have different tolerances to water acidification. Similarly, many plant species are indicative of eutrophication.

A wide range of techniques, including standard survey procedures and modelling software for analysis of the results, are now available for the pollution manager, and these are proving very robust for a wide range of purposes. It is not possible to examine this area in detail here, but comments will be made on biological and chemical monitoring at appropriate points elsewhere in this chapter.

Pollution from metals

There are a number of sources of pollution by *heavy metals* (metals with high atomic weights). These are:

1   *Natural geological weathering.* This is the chief source of the low background levels in pristine waters. However, in a few locations where surface waters

receive water from metal-rich ores, very high natural concentrations can occur.

2   *The processing of metal ores.* This is a major source of water pollution in areas remote from urbanisation, and can result in concentrations so high that all life in the affected rivers may be killed (Kelly, 1988). A particularly problematic source of pollution is abandoned mines, which fill with water and then spill into nearby rivers. This is expensive to treat, and usually the liability for the costs is difficult to determine. The recent closure of many mines in Britain is likely to result in this problem becoming increasingly important.

3   *Sewage.* Animals do excrete some metals (especially zinc) and many sewers contain other small sources of metals. Inefficient treatment of sewage can lead to significant pollution by metals.

4   *Industrial and other uses of metals.* This includes lead from combustion of petrol and resulting run-off from motorways.

5   *Leachate from landfill.* Many landfill sites contain metal waste which slowly corrodes and becomes soluble. If such leachate enters surface waters then significant concentrations of metals can result (along with other pollutants) (Cheung *et al.*, 1993).

There are not many cases of acute toxicity to humans arising from metal pollution in water. One example occurred in Taiwan, where high levels of arsenic in water were reported following volcanic activity, and this led to blood diseases and skin cancer. Perhaps the best-known example of metal pollution affecting humans is that of lead from the water distribution system itself. In many countries, lead pipes have historically been used to convey water to domestic users. If the waters are acidic (soft waters) then the lead will slowly dissolve. Lead poisoning can result, and most countries have programmes under way to replace these pipes. As with many pollutants, the sections of the population most vulnerable to metal poisoning are the old, young and sick.

The effects of metals on ecosystems can be severe. Some metals are essential as micronutrients for plant growth, e.g. copper, iron, manganese, molybdenum and zinc. However, in high concentrations they can produce toxicity effects in plants and animals. Other metals, e.g. cadmium, lead and mercury, are not required for metabolism, and are often highly toxic.

Some plants seem highly resistant to many metals. This is particularly so for *bryophytes*. The cell walls of these plants have a very large *cation exchange capacity*. Thus they can accumulate cations, such as heavy metals. In pristine waters, metal cation nutrients such as calcium and magnesium may predominate. However, in polluted waters bryophytes will accumulate many metal pollutants. This makes them useful biomonitors of this type of pollution, reflecting the general state of water quality rather than an instantaneous measure that would be made by direct water sampling (Glime, 1992). This is most useful as a character for monitoring pollution in waters contaminated by a single metal, or for monitoring in remote areas where repeated collection of water samples may be difficult. However, when more than one metal is present in high concentrations, they may compete with each other for exchange sites in the cell walls. Ruhling and Tyler (1970) noted that some metals are

PLATE 4.1 The problems of heavy-metal contamination of soils and waters from industrial activity may be enormous. The lower Swansea valley, shown here, was contaminated with a large number of different metals from a variety of sources and exten-sive remediation was necessary to reduce the toxicity.

Photograph: Andrew Farmer.

more likely than others to bind to aquatic bryophytes. The order was: Cu, Pb > Ni > Co > Zn, Mn. The presence of high copper levels therefore, means that there would be an underaccumulation of zinc. Examples of the very high concentrations found in some bryophytes are given in Table 4.5.

TABLE 4.5 Concentrations of metals found in the tissues of aquatic bryophytes

Metal	Bryophyte species	Concentration (% of dry weight)	Reference
Copper	Drepanocladus fluitans	0.035	Erdman and Modreski (1984)
Iron	Jungermannia vulcanicola	5.0–13.0	Satake et al. (1989a)
Lead	Scapania undulata	0.7–2.4	Satake et al. (1989b)
Mercury	J. vulcanicola	1.3	Satake et al. (1983)

There is a wide range in the sensitivity of plants to metals and this may vary depending on the metal. For example, the moss *Fontinalis antipyretica* is known to accumulate large concentrations of many metals, but is sensitive to copper pollution. Many flowering plants are sensitive, and there is widespread variation among algae, with blue-green algae usually being the most resistant. In general, significant metal pollution will reduce productivity in freshwaters and this will lead to general changes in ecosystem function.

Organic pollution

The discharge of plant and animal material (organic pollution), which provides a carbon source for the growth of micro-organisms and higher plants and animals, can be a serious pollution problem. The measure of organic pollution is the 'biological oxygen demand' (BOD) of the water. The more material available for decay, the higher the oxygen use by the micro-organisms and hence the higher the BOD. There are many sources of organic enrichment in freshwaters. These include sewage treatment works, agriculture, fish farms, landfill and general run-off from the land (e.g. after flooding). The relative 'strength' of the different sources of organic pollution is given in Table 4.6 by the measure of their BOD.

A common feature of severe organic pollution is the development of sewage fungus. This is a slimy, furry mat in the water, which can break off and float downstream. Sewage fungus is not, in fact, a fungus, or even a single species of organism. It is a colony of bacteria, algae, fungi and zooplankton. However, the dominant species present is the bacterium *Sphaerotilus natans*. The fungus can trap other material in the water and the slow decay of the mats can produce an unpleasant odour, which can even lead to a public nuisance.

The growth of sewage fungus and other bacterial growths usually eliminates algal growth for some distance away from organic discharges. A number of *macrophytes*

TABLE 4.6 The biological oxygen demand for different sources of organic pollution

Pollutant source	Biological oxygen demand (mg/l)
Milk	140,000
Silage effluent	30,000–80,000
Pig slurry	20,000–30,000
Cattle slurry	10,000–20,000
Yard washings	1,000–2,000
Domestic sewage (raw)	300–400
Domestic sewage (treated)	20–60

Sources: MAFF (1991) and SOAFD (1992)

Note: Due to variations in the quality of some sources, ranges may be given.

are tolerant to this pollution and their presence along with few other species may indicate a problem. These include *Enteromorpha*, *Potamogeton crispus* and *Sparganium* spp. However, one species, *Potamogeton pectinatus*, actually seems to be positively promoted by organic pollution and will occur in polluted areas outside its native range. It is, therefore, a useful biological indicator.

Bacteria are not the only organisms that can feed on decaying organic matter – invertebrates may also respond positively. In severely polluted water, however, invertebrates are likely to be eliminated. The most tolerant species are the small, red *Tubifex tubifex* worms, along with a few other worms and the drone fly *Eristalis tenax*. As the pollution declines, the species appearing include *Chironomus*, various leeches and molluscs, and finally the stoneflies and mayflies.

Suspended solids

The discharge of suspended solids often occurs alongside organic pollution, and it is predominantly caused by activities that work the land, e.g. farming and the minerals industry.

Suspended solids have a physical effect on the aquatic environment – they prevent light penetration. This reduces the light available for photosynthesis by macrophytes and algae. This will result either in the loss of such plants, or a reduction in the depth to which they can colonise. In any case, a decline in the primary productivity of the water body will take place.

Tolerant plant species are either those that avoid the shade (e.g. they float or are emergent), are highly tolerant of shade, or can grow quickly to escape it. Such species include *Lemna minor*, water lilies and *Sagittaria* spp., while less tolerant species include *Potamogeton perfoliatus* and *Elodea canadensis*.

Invertebrate communities will also change. A decline in the number of submerged plants will reduce the populations dependent on these, and the deposition of suspended solids will affect any species dependent on bare sand and gravel these being replaced by those that burrow. Stoneflies and mayflies are, therefore, particularly sensitive, while tubificid worms and chironomids are tolerant. Suspended solids can also lead to reductions in fish populations, by either affecting gill function or by covering spawning areas. Those fish species that require very clean, clear water and spawning gravel, such as trout, are highly sensitive to pollution by suspended solids.

CASE STUDY: TEMPORARY CONTROL OF SEDIMENTS IN STREAMS FROM SEDIMENT SOURCES

Many sources of sediment in streams and rivers are temporary, for example from construction operations. The quantity of sediment may be large, but only short-term measures need to be taken to control it. Some methods of control (e.g. diverting streams) can be very expensive and uneconomic. The New York State Electric Company undertakes many operations near stream systems and therefore has developed a control system that can be placed in streams; this traps the sediment for the period of operation and is easily removed after use. The method involves sediment mats (Anon., 1996). These consist of a lower layer of hessian, a middle layer of wood-wool and an upper layer of jute mesh, and are 1.2 m × 3 m. These mats are fixed on to the stream beds below the sediment source. The sediment moves through the outer jute mesh, but cannot pass through the hessian layer, and so it accumulates in the wood-wool. Tests have shown they trap about 80 per cent of the sediment, and each mat will hold up to 200 kg. After the sediment source ceases (or when the traps are full), the traps can be easily removed and rolled up. As they are biodegradable they can readily be recycled to be used in operations such as stream bank stabilisation.

CASE STUDY: TREATMENT OF RUN-OFF FROM ROADS

During heavy rainfall, roads can produce significant quantities of polluted water. Some of this enters storm sewers and is treated. The rest will reach surrounding soils, wetlands and surface waters. Such water is polluted with a range of substances arising from vehicle use, e.g. organic pollutants, heavy metals, etc. (Dobson, 1991). A range of options are open for the control of such pollutants (see SEPA, 1996). Three general approaches may be taken – reducing the quantity of run-off, slowing its velocity, or treating it. Examples of each include:

1   Reducing run-off (all of these not only reduce run-off, but also help filter out pollutants):
    (a)   *Infiltration trenches.* These form a stone-filled reservoir alongside a road, which retains much of the run-off; this can then filter into the subsoil.

(b) *Porous pavements.* These may use particular forms of asphalt, which allow water penetration to underlying gravel and soil layers.

(c) *Infiltration basins.* These are impoundments designed to store storm water and allow much of it to enter the underlying soil.

2   Slowing velocity (slowing velocity will increase the rate of sedimentation of suspended solids):

(a) *French drains.* These are gravel-filled ditches containing a drainpipe system. The gravel reduces water velocity and helps to remove pollutants.

(b) *Swales.* These are wide surface depressions alongside roads that are vegetated. The large surface area and surface roughness reduces water velocity and also helps to filter some pollutants.

3   Treatment:

(a) *Filter strips.* These are areas of vegetated land designed to hold surface run-off and allow its seepage into the soil.

(b) *Detention ponds.* These may either normally be dry or wet, but act to allow suspended solids to settle out.

Perry and McIntyre (1986) compared different methods of treating motorway run-off. Two of these are presented here. The first is a simple sedimentation tank. The run-off enters a large tank, where particulates can settle out. The second is a more complex lagoon system with layers of gravel and sand, partly overgrown by vegetation and supported by a thick plastic sheet. This bears a similarity to the use of reed-beds for treating effluent (see p. 128). Table 4.7 shows the effectiveness for some of the major pollutants.

TABLE 4.7   Percentage removal of pollutants by two treatment methods compared with untreated motorway run-off

Pollutant	Sedimentation tank	Lagoon
Suspended solids	52	92
Total lead	40	90
Total zinc	47	76
Oil	30	70

This demonstrates that even a simple sedimentation tank can remove significant quantities of pollutants. However, very high efficiencies can be obtained by more complex systems. It is not clear, however, how such complex systems behave in the long term after many years of filtering pollutants. Their efficiency would certainly be reduced and their replacement would be more costly than cleaning a sedimentation tank.

Nitrate pollution

There has been increasing concern about the effects on human health of nitrate in drinking water (Dudley, 1990). Of most concern is an effect which is most prominent in infants and, to some extent, the elderly. In the body the nitrate is reduced to nitrite. The nitrite is readily absorbed through the stomach and it then reacts with the haemoglobin in the blood, forming methaemoglobin. If about a third of the haemoglobin reacts in this way a condition known as methaemoglobinemia results which is a form of *hypoxia*. Fatalities occur when about half of the haemoglobin has reacted with the nitrite. Since 1945 about 2,000 cases of methaemoglobemia have occurred world-wide. Seventeen of these have been in Britain, with only one fatality. Babies are considered to be at greatest risk, and nursing mothers are advised to use bottled water in areas where high nitrate levels occur. The UK water company, Anglian Water has, for example, two mobile water bottling plants ready to supply bottled water should nitrate levels exceed 100 mg/l. There is also concern that nitrate in drinking water can cause cancer of the stomach. However, this relationship has proved more difficult to demonstrate.

Nitrate levels have increased considerably in many rivers and groundwaters over recent decades. Table 4.8 provides a few examples of this.

TABLE 4.8   Changes in nitrate concentrations in four English rivers from 1928 to 1976

River (and sampling location)	Annual average nitrate concentration mg/l		
	1928	*1960*	*1976*
Great Ouse (Offord)	—	2.0	8.0
Lee (Chingford)	3.0	5.0	12.0
Stour (Langford)	3.0	4.0	7.5
Thames (Hampton)	2.5	4.5	8.0

Source: Marsh (1980)

The most important source of nitrate is agricultural applications of nitrogen fertilisers, although additional important sources include sewage treatment works and industry. Drinking water is abstracted from both surface waters and groundwater and so pollutant impacts on both of these sources need to be addressed.

Regulation: nitrate vulnerable zones

In areas where nitrate pollution of drinking water is a problem, reductions in input from agriculture must be achieved. The UK set up a scheme for the identification of 'nitrate sensitive areas' (NSAs), where groundwaters were potentially threatened by nitrate. Within these areas farmers were offered a *voluntary* scheme whereby they could receive payments for less-polluting farming practices (e.g. less fertiliser use or reversion to grasslands).

However, since the creation of NSAs, the EU has agreed its own strategies to reduce nitrate pollution under the Nitrates Directive, i.e. via the adoption of 'nitrate vulnerable zones' (NVZs) (DoE, 1993). Under this system, groundwaters and surface waters are both protected from nitrates, either as a drinking water issue or for eutrophication. Within an NVZ, controls are *mandatory* and farmers receive no compensation. The controls on eutrophication are only enforceable where nitrogen is known to be the limiting nutrient. In most UK rivers, however, phosphorus is limiting, so the controls have little benefit for this problem. The mandatory nature of the NVZs has caused significant opposition from farming organisations, who challenge both the efficacy of some of the proposed controls and the accuracy of the nitrate drinking water standards themselves. This is a good example of where pollution management requires very accurate information for decision-making, because of the significant economic consequences that there may be for a large number of individuals.

In order to reduce nitrate pollution, the UK government has recommended the following actions in order to reduce impacts:

◆ Reduce the application rates of organic manures to only those which are necessary and apply them when the nitrogen can be used by the crops (e.g. do not apply in autumn or early winter).

◆ Accurately calculate the required amount of inorganic fertiliser required by assessing current soil status and crop requirements, and apply only this amount.

◆ Try to avoid leaving soil bare for long periods, as this increases the leaching of nitrate.

◆ Incorporating crop residues (e.g. straw) in soils that are low in nitrogen helps to reduce leaching, but incorporating residues high in nitrogen (e.g. beans) should be avoided.

◆ For grasslands, a reduction in grazing pressure (especially in autumn and winter) will significantly reduce nitrate loss.

◆ Ploughing up grassland causes a flush of nitrate. This should only be done when necessary.

◆ Irrigation measures need to be accurate. Too much water will increase the leaching of the nitrate, but too little will mean that added nitrogen will not be utilised by the crop.

Eutrophication

Eutrophication can be defined as adverse ecosystem responses to excess nutrient supply. For freshwaters the nutrients of most concern are phosphates and nitrates, and their impacts are extremely varied and potentially very destructive (Harper, 1991;

Haslam, 1990; Hellawell, 1988). The two most common sources of nutrient enrichment are agriculture and domestic sewage. The former can supply large inputs of either nitrates or phosphates, or both, depending on the type of operation being undertaken. Primary sources of agricultural pollution arise from the direct infiltration of fertilisers applied to fields and also from the leachate from silage silos or animal slurry. Domestic sewage also supplies both nutrients although concern is most often expressed over phosphate discharges. Eutrophication can also occur due to run-off from landfill sites, industrial discharges and the operation of fish farms.

Eutrophication is probably the most serious threat to freshwaters on a global scale, and while the problem is being tackled in some locations, the threat continues to grow in many countries. This is especially so for those locations where there is a significant population increase and where urbanisation is occurring.

The Organisation for Economic Cooperation and Development (OECD) has produced a classification of the trophic status of freshwaters, based on the total phosphorus concentration in the water column. This is detailed in Table 4.9. The phosphorus concentration can also be related to other characters in the water column such as chlorophyll (i.e. algal biomass), water clarity and oxygen saturation (measure of biological activity). However, these latter measures are not always a good guide to trophic status. For example, the OECD suggest that a *Secchi depth* of 1.5–3.0 m is an indication of eutrophication. However, most oligotrophic Scottish lochs have such a Secchi depth, but the reduced visibility is due to dissolved organic acids derived from the surrounding peaty soils and has nothing to do with eutrophication.

TABLE 4.9 OECD categories for trophic status based on the total phosphorus concentration in the water column

OECD category	Total phosphorus concentration ($\mu g/l$)
Ultra-oligotrophic	<4
Oligotrophic	4–10
Mesotrophic	10–50
Eutrophic	50–100
Hypertrophic	>100

It is very important to note that water bodies that fit the description of eutrophic or hypertrophic are not necessarily polluted. There are many naturally highly productive water bodies, with extensive algal, plant and fish growth. In first assessing trophic problems in a lake or river, it is important to determine whether any changes have occurred, hopefully by examining historic data on water chemistry or biology. Alternatively, it is possible to take sediment cores in lakes and examine changes in the diatom (unicellular algae) populations over many decades. These can be used to model changes in trophic status over time, as individual species are characteristic of different water quality conditions. In principle, the technique is similar to that used to examine changes in acidification status, and so the technique will be described in more detail later in this chapter.

The characteristics of waters in each OECD category can broadly be defined according to its biological and chemical criteria. These are given in Table 4.10.

TABLE 4.10 Characteristics of freshwaters of different trophic status based on OECD categories

Character	Ultra–oligotrophic	Oligotrophic	Mesotrophic	Eutrophic	Hypertrophic
Biomass	Very low	Low	Medium	High	Very high
Green and/or blue-green algal fraction	Low	Low	Variable	High	Very high
Macrophytes	Low	Low	Variable	High or low	Low
Productivity	Very low	Low	Medium	High	High, unstable
Oxygen dynamics: Epilimnion (upper water layer)	Normally saturated	Normally saturated	Variably saturated	Often over-saturated	Very unstable, often no oxygen
Hypolimnion (deep water layer)	Normally saturated	Normally saturated	Variably under-saturated	Under-saturated to complete lack	Very unstable, often no oxygen

Source: Meybeck *et al.* (1989)

The growth of freshwater plants is primarily limited either by lack of light (in deep or turbid water) or by nutrient supply. Many plants are adapted to low nutrient conditions. However, if the nutrient limitation is overcome by the addition of pollutants, then the more aggressive species will respond and can come to dominate the freshwater ecosystem. The following impacts are likely to occur:

1   The first impact of nutrient enrichment to freshwaters is a bloom of phytoplankton. These microscopic plants are able to multiply very rapidly if all their nutrient needs are met, and water bodies can very rapidly turn into a green 'soup'. Alternatively, the blooms may appear as a 'scum' on the water, particularly around the edge of lakes. Such blooms reduce the light available to higher plants in the lake or river. In many instances, the bloom is temporary,

as the zooplankton populations that graze on the phytoplankton also increase in numbers. However, continual excessive nutrient addition can maintain a bloom. In Europe and North America such blooms often occur in spring and autumn, and this reflects the rainfall patterns, when run-off from agricultural land containing nutrients occurs and temperatures are still adequate to promote algal growth. The appearance of a bloom is a good indication that a problem exists, and its source should be investigated and rectified.

2   Excessive growth of aggressive higher plants. Submerged species, such as *Hydrilla verticilata* in Florida or *Myriophyllum spicatum* and *Potamogeton pectinatus* in Europe, directly outcompete other species within the water body. Equally damaging are floating species, e.g. *Lemna* species in temperate regions or *Eichornia crassipes* in tropical ecosystems. These form a complete blanket on the water surface and so deprive submerged species of light. The reduction in the growth of submerged plants will reduce the production of oxygen in the water bodies. Oxygen-deprived waters (anoxic waters) are unable to support fish.

3   Excessive growth of filamentous algae and *epiphytes*. Epiphytes grow on the leaf surfaces of other plants and the growth of these aquatic algae can prevent light reaching the plants on which they grow. This is a particularly serious problem when very nutrient-poor lakes (oligotrophic lakes) are polluted, as the plants characteristic of these lakes are often slow-growing and are easily smothered. Filamentous algae can form extensive mats within a water body, and can smother all the other plants that it contains. Blue-green algae (more closely related to bacteria than other algae) may form these mats. These can be a particular problem, as they also release toxins that can lead to fish kills and may even pose a threat to humans and their domestic animals while swimming in the waters that contain these toxins. Even though these blue-green mats may not cover a lake, their presence in sufficient quantities will lead to warnings being issued about bathing in such lakes.

Managing eutrophication

The common principles of waste management applicable to other forms of waste also apply to managing eutrophication, i.e. prioritising waste prevention, then its minimisation, followed by recycling, treatment and then management *in situ*.

The reduction in the production of nutrients entering the waste stream is of primary importance. This is the cheapest way to deal with pollution. The case study examining the reduction in inputs from agricultural silage production illustrates that basic housekeeping measures can be very effective. For domestic waste production other measures are necessary. For example, although there has been a significant reduction in the amount of phosphorus derived from the use of domestic and industrial detergents, they still account for about a quarter of the phosphorus produced by each person in Western Europe. Some of the reduction required can be voluntary, and awareness of the issues has certainly resulted in the greater use of phosphate-free products. However, there may also be a greater role for government (including at a European Union level) regulation.

Farm-derived nutrients enter watercourses and can only be managed further *in situ* (NCC, 1991; NRA, 1992). However, those derived from many domestic and industrial sources enter sewers and are amenable to further treatment. Such treatment is still not widespread enough to prevent many problems. For example, the case study on the impacts of eutrophication on nature conservation sites in England shows that insufficient treatment at sewage works is the biggest source of the problem at such sites.

Sewage treatment works may operate a variety of treatment procedures. At the very least they provide sedimentation tanks and filter beds to remove suspended solids. However, further treatment to remove dissolved nutrients (advanced waste-water treatment) is not always provided, particularly at sewage treatment works serving smaller communities. A range of biological and chemical options are available to remove phosphorus and nitrogen, and the actual method used depends on cost requirements and local circumstances.

The most widely used method of phosphorus removal is by chemical treatment with either iron sulphate or aluminium sulphate. This results in the production of insoluble iron or aluminium phosphate, which precipitates out as a floc in the water and which is readily removed. This particular treatment can be used at most stages in the treatment of wastewater, and usually results in the removal of about 90 per cent of the phosphate in the water column.

Treatment for nitrogen can also employ a chemical method. In this case, the primary ion removed is ammonium (which is toxic to fish). This can be removed by the use of ion-exchange resins, which replace the ammonium with a non-toxic metal cation (e.g. potassium or calcium). However, although 99 per cent of the ammonium can be removed this way, this type of treatment is expensive, and biological methods are usually favoured. The biological methods rely on microbial action and are a two-stage process. In the first stage, bacterial action oxidises the ammonium to nitrate. This nitrification process requires good aeration and usually involves either the use of aerated gravel beds or activated sludge. The second stage involves the denitrification of the nitrate to nitrogen gas, which is lost to the atmosphere and is harmless. This stage requires more anaerobic conditions and the addition of a carbon source to provide energy for the microbial metabolism. The two processes can take place in one continuous, purpose-built construction.

Other biological methods can be employed. These remove both nitrogen and phosphorus from wastewaters. The simplest method is to cultivate fast-growing plants in the water and, as the nitrogen and phosphorus is taken up into the plant tissues, they can be harvested. Floating plants are best for this method. Such plants are not limited by nutrient supply, carbon supply, water or light. In tropical and subtropical climates their growth is extremely rapid. One species commonly used is the water hyacinth, *Eichornia crassipes*. This species is an extremely aggressive weed in open waters in the tropics of both Africa and the Americas. However, these characters make it ideal for treatment of wastewater. Disposal of the harvested material can be a problem, but composting as green manure, or combustion and utilisation of the ash as fertiliser are options, providing the wastewaters are not significantly contaminated with heavy metals or other toxic substances that may also be absorbed by the plants. In this case, landfill or careful control of combustion conditions may be needed.

In Europe, the construction of artificial reed-beds for water treatment is becoming more common. The use of the common reed (*Phragmites*) takes advantage of a number of the characters of such beds. The reeds actively aerate the sediment in which they grow, as they pass oxygen from the leaves down to their root systems. This provides the aerobic conditions for nitrification of ammonia. However, the sediments also act as a filter for suspended solids and absorption of metals. The often high iron content of these sediments also helps to remove phosphorus. Large reed-beds also have the advantage of providing a useful habitat for wildlife species and this could be particularly important in urban areas where such habitat is limited. However, the efficiency of such beds does decline with age and a considerable area of ground needs to be available for their construction.

A recent survey (Haycock, 1996) of the efficiency of 285 European reed-beds for pollution control has revealed a number of problems. Generally, such beds perform well in reducing BOD (the best may produce a 78 per cent reduction) and suspended solids. However, most are poor at reducing ammonium (with an average reduction of only 17 per cent) and total nitrogen (average reduction of 43 per cent) and phosphate (average reduction of 65 per cent). The survey concluded that the problems in the ability to reduce nutrient pollutants was due to the way that the constructed wetlands work. It seems that they only begin to reduce the nutrient loading *after* the BOD has been reduced significantly, but for many their size prevents this. It is generally assumed that a wetland can control BOD at the rate of 12–15 g/m^2/day (as one person produces 60 g/day, this requires 4–5 m^2 per person). However, construction limited to these parameters will fail to reduce other pollutants to a significant level. The survey recommended that, in order to control ammonium, wetlands should be constructed on the basis of 6–7 g BOD per m^2/day. If total nitrogen is included this should be reduced to 4–5 g BOD per m^2/day and for phosphate only 2–3 g BOD per m^2/day.

Many waters are already eutrophic and need to be managed as they are. In other instances, the reduction of inputs may either be technically difficult or be less politically desirable (e.g. in restricting agricultural practices). A number of techniques have been developed for use in water bodies (almost always standing waters) to remove nutrients.

Techniques used for treatment in sewage treatment works can also be applied to lakes. For example, the addition of iron sulphate to remove phosphate can be undertaken. For the last few years iron sulphate has been added to Rutland Water in Leicestershire, England. This reservoir has great amenity value, but the high nutrient levels have resulted in a number of blue-green algal blooms. The reservoir is also a statutory nature conservation site noted for its bird populations. Blue-green algae can produce toxic substances and are dangerous to humans and their animals. Treatment with iron sulphate has resulted in the reduction of phosphate and blue-green algal levels. However, the result is a floc of precipitated iron phosphate. This covers large parts of the reservoir and so causes impacts on invertebrate populations. Over this time bird populations have also fallen. It is important, therefore, to note that treating one aspect of a eutrophication problem can result in others being created. Priorities must always be determined for any water body before management options are acted upon.

An interesting mechanical technique for treatment of lakes is to aerate them. One result of eutrophication is to reduce the dissolved oxygen in the deeper parts of the lake (the hypolimnion). Aeration enables full decomposition of dead plant material to take place and nitrification processes to occur. More aerobic conditions also allow for precipitation of iron phosphate from iron naturally present in the deeper water and sediments.

Perhaps the most drastic mechanical measure is to remove the sediments from a lake, which is usually done using a large suction device, which removes the sediment to barges or to the banks of the water body. Many eutrophic lakes are unable to recover due to continual resuspension of nutrients from the sediments. Such resuspension can be for various reasons, e.g. the action of fish (see below), by motor boats (e.g. in the Norfolk Broads, England) or by natural processes. Thus if recovery is required, this must be overcome. Given the expense of the operation and problems of disposal of the sediments, this is only practical where the water body is of very high amenity or conservation value.

Biomanipulation is also an option. At the simplest level if fish are present which actively stir the sediment (e.g. carp) these can be removed. However, there are also more subtle measures aimed at altering the whole ecosystem functioning of a lake. For example, if algal phytoplankton blooms are a major problem for a lake, then measures can be taken to encourage the production of the zooplankton that eat these phytoplankton. This might either be done by attacking the populations of the fish that eat the zooplankton or by boosting the populations of predatory fish that feed on the zooplankton feeders. The procedures for such measures are now well established. These measures may result in some changes to water quality, but are primarily aimed at relieving some of the symptoms of eutrophication.

CASE STUDY: CONTROLLING WATER POLLUTION FROM FARM SILAGE PRODUCTION

In 1991 the UK Ministry of Agriculture, Fisheries and Food (MAFF) issued guidelines for the protection of water from a range of pollutants arising from agriculture. This case study outlines those guidelines applying to the production of silage.

Silage production is subject to statutory regulation, The Control of Pollution (Silage, Slurry and Agricultural Fuel Oil) Regulations 1991. These regulations allowed farmers to use existing silos and other methods (if agreed). However, for new production, new standards of silo construction were required, or regulations for the use of sealed plastic bags as silage sources.

New silos included the following conditions:

1   The floors of silos should be impermeable and corrosion resistant, with channels around it to collect effluent.

2   If the silo has walls, the base should extend beyond these. These should also be resistant to corrosion.

3   The effluent should be channelled to a tank, which could hold at least 20 litres for each 1 m^3 of silo space and be resistant to corrosion by acid.

4   No part of the silo or tank should be within 10 m of a watercourse.

5   In constructing a silo, any drains under the site should be moved to at least 10 m from the silo.

6   The floor of the silo should slope towards the channels in the silo, to reduce pressure from the liquid on the walls.

7   Regular inspections are necessary to ensure that cracks, etc., are identified and repaired.

8   It has been recommended that crops for silage are allowed to wilt before use to reduce the amount of water available for effluent production.

For the production of silage by bales, the same recommendation applies to the crops. In addition, the 1991 regulations required that the bales should be stored at least 10 m from a watercourse and they should not be opened within 10 m of a watercourse or field drain.

CASE STUDY: USE OF STRAW TO MANAGE ALGAL BLOOMS

Algal blooms resulting from eutrophication are best controlled by reducing the source of the pollution. However, this may not always be possible, and there is a simple technique that has been shown to be effective in treating water bodies with algal problems. The method involves the use of straw (usually from barley).

Application of straw to a water body eventually leads to the decomposition of the straw. This produces a chemical substance (the identity of which is not yet known), which acts to prevent unicellular and filamentous algal growth and reproduction, although it does not kill the algae (Gibson *et al.*, 1990; Ridge and Barrett, 1992). The chemical does not affect higher plants or animals, although excessive application of straw and its decomposition could lead to deoxygenation. In fact, Everall and Lees (1996) have reported that its use in reservoirs does provide some additional benefits by the creation of secondary nesting areas for waterfowl and substrates for colonisation by beneficial invertebrates. In general the use of straw is a cheap and safe management tool.

The rate of decomposition will, of course, vary with conditions. In cool waters it may take up to eight weeks to begin to be effective, but in the summer, in warm waters, this can be achieved in about a week. A good oxygen supply is also necessary in order to encourage decomposition. Activity remains as long as there is straw to decompose, and one dose can be sufficient for an entire growing season. However, the chemical can be absorbed by sediments, and therefore for waters with high suspended solids or which are regularly stirred, larger and more frequent doses may be necessary. Once activity has begun, algal blooms will begin to disappear. This may take a few weeks for unicellular algae, but longer for filamentous species or in sites where very difficult problems occur.

Straw can be applied to both lakes and rivers. In lakes, straw should be applied loose to increase aeration, but in rivers this would be easily lost and bales should be anchored into the water. Newman (1994) suggests the following application rates of straw:

Lakes and ponds: 2.5 g/m^2 of water surface. In places with a severe problem, initial high doses of up to 50 g/m^2 may be needed.

Rivers: About 20 kg (one bale) every 30–50 m of distance along the river.

CASE STUDY: IMPACT OF EUTROPHICATION ON THE NATURE CONSERVATION INTEREST OF FRESHWATER SITES IN ENGLAND

On a more detailed scale Carvalho and Moss (1995) undertook an examination of the status and reasons for eutrophication on freshwaters at Sites of Special Scientific Interest (SSSIs) in England. SSSIs are statutorily protected areas and are identified for their nature conservation importance. Obviously, eutrophication can lead to a decline in the conservation interest of a site, eventually to a point where that interest is lost. Existing data for 102 SSSIs were examined to determine whether eutrophication had taken place. The initial trophic status of the water body was classified (see Table 4.11) according to the OECD categories in Table 4.9.

TABLE 4.11   The number of SSSIs in England showing symptoms of eutrophication

OECD category	*Number of SSSIs examined*	*Number with symptoms of eutrophication (%)*
Oligotrophic	2	0 (0)
Mesotrophic	13	9 (69)
Eutrophic	13	10 (77)
Hypertrophic	65	57 (88)
Unknown	9	9 (100)
Total	**102**	85 (84)

Source: Carvalho and Moss (1995)

Note: Figures in brackets indicate percentage of SSSIs affected for each category.

Carvalho and Moss (1995) undertook a more detailed examination of seventy-nine of the eighty-five SSSIs with eutrophication symptoms to determine the source of the problem. The results are shown in Table 4.12.

Having identified the main causes of eutrophication, Carvalho and Moss (1995) proposed a series of primary management requirements for action to restore the sites and prevent further damage. These are detailed in Table 4.13.

TABLE 4.12 The primary cause of eutrophication in seventy-nine SSSIs in England

Cause of eutrophication	Number of SSSIs affected (%)
Main sewage effluent	35 (44)
Possible main sewage effluent	5 (6)
Fish community	15 (19)
Possible fish community problem	6 (8)
Farm animal wastes	7 (9)
Possible farm animal wastes	7 (9)
Septic tanks	3 (4)
Possible septic tank problem	1 (1)
Landfill effluent	1 (1)
Possible landfill effluent problem	1 (1)
Migratory gull roost	1 (1)
Sediment release	1 (1)
Complex reasons, none predominant	5 (6)
Unknown	6 (8)

Source: After Carvalho and Moss (1995)

Note: Figures in brackets are the percentages affected for each category.
The figures total more than seventy-nine (100 per cent) as more than one cause may affect each site.

While some of the sites affected by sewage discharges (the main cause of eutrophication) should be improved by planned modifications to sewage treatment works, many of those affected by sewage treatment works serving populations of under 100,000 will not be improved under the Urban Waste Water Directive. Carvalho and Moss (1995) suggest that the main UK government motivation in applying the Directive in the UK is in improving water supply, rather than wider environmental impacts. For example, no near-pristine sites are currently on the government list for action and yet these would benefit greatly from improvements.

Biomanipulation is also needed in some instances. For example, the action of carp and bream in stirring the sediment leads to eutrophication, and in some instances these fish need removal. However, such action also needs to take account of the interests of anglers.

As a means of strategic management of phosphate pollution in freshwaters of nature conservation interest, the water pollution regulator in England, the National Rivers Authority (predecessors of the Environment Agency), and the statutory nature conservation agency, English Nature, developed target Special Ecosystem levels for phosphate concentrations in rivers for different ecosystem types. These levels are present in Table 4.14. It is important to note that phosphate concentrations in sewage

TABLE 4.13 Primary management requirements for the seventy-nine SSSIs affected by eutrophication

Management required	Number of SSSIs affected (%)
Sewage effluent diversion or phosphorus removal	30 (38)
Possible need for effluent diversion or phosphorus removal	4 (5)
Fish community manipulation	18 (23)
Possible need for fish community manipulation	1 (1)
Need for control over fish stocking	6 (8)
Buffer zones, wetland enhancement	12 (15)
Diversion of farm stock wastes	2 (3)
Possible need for diversion of stock wastes	5 (6)
Improvement in septic tank function	3 (4)
Reroute storm drain	1 (1)
Flood containment	1 (1)
Sediment removal	2 (3)
Diversion of landfill leachate	2 (3)
Salinity reduction	1 (1)
Continue or establish monitoring	16 (20)
Further studies required	35 (44)

Source: After Carvalho and Moss (1995)

Note: Figures in brackets are the percentage for each category. The number (and percentage) total more than seventy-nine (100 per cent) as more than one management requirement may be needed for any one site

TABLE 4.14 The Special Ecosystem target levels for phosphate in freshwaters

River type	Orthophosphate concentration (mg/l)
Upland rivers (maximum)	0.02
Chalk, limestone, hard sandstone (maximum)	0.06
Clay, alluvial lowland rivers (maximum)	0.10
Heavily enriched rivers (initial target)	0.20

without secondary treatment are about 10 mg/l, while after secondary treatment they are reduced to about 1 mg/l. Thus with normal dilution in rivers, normal secondary treatment would be sufficient to achieve the standards needed for England's sensitive freshwater ecosystems.

English Nature (1994) has estimated that of the twenty *river* SSSIs and the Norfolk Broads threatened by eutrophication, additional secondary treatment at thirty-five sewage treatment works would be needed to provide protection and meet the Special Ecosystem target levels for phosphate. The cost of installing phosphate removal at a sewage treatment works for a town of 10,000 people is about £100,000, with annual running costs of about £19,000. The costs of different-sized works necessary to clean up the thirty-five sewage treatment works is estimated at less than £10 million, including running costs for the first five years. While this seems a large sum of money, the water industry's current investment programme is about £23 billion. Thus dealing with the nature conservation impact would not amount to a large proportion of the investment budget. However, such work does need to be balanced against other measures, such as meeting EU directives for bathing and drinking water and reducing loss of water via leakage from pipes, which became a significant political issue during the summer drought in Britain in 1995.

CASE STUDY: MANAGEMENT TO RESTORE A SHALLOW EUTROPHIC LAKE, NORFOLK BROADS, ENGLAND

From 1982 to the present day, work has been undertaken to restore the ecosystem of Cockshoot Broad in Norfolk. Like many of the Broads, the lake is shallow (about 1 m depth) and was formed in the Middle Ages following flooding of old peat workings. For many centuries the Broads have supported rich plant and fish life. However, during recent decades, they have become progressively eutrophic due to excessive phosphorus discharges from sewage. At Cockshoot Broad efforts have been undertaken to remedy this, and these have been described by Moss *et al.* (1986) and Moss *et al.* (1996).

The researchers describe a series of phases in the restoration management, with each having its own impact. The changes in phosphorus, chlorophyll and *Daphnia* are shown in Figure 4.1. The stages were:

1   Prior to isolation the Broad was fed by the River Bure. This was a nutrient-rich source, high in nitrogen and phosphorus originating from poorly treated sewage effluent. The lake had high total phosphorus levels, high chlorophyll levels and low levels of zooplankton such as *Daphnia*, and very poor macrophyte communiti;s.

2   In 1982 the Broad was isolated from the River Bure, and its source of water changed to a drainage stream with much lower nutrient loading. As the nutrients already received into the Broad had accumulated in the sediments, a metre depth of this was also removed at this time by suction dredging. The outflow was also protected against backflow. Following this treatment, total phosphorus and chlorophyll levels fell rapidly and *macrophytes* colonised one part of the lake. Initial fish populations were very low, probably due to the original disturbance from the sediment removal.

3   In 1985 fish populations began to recover and zooplankton populations fell.

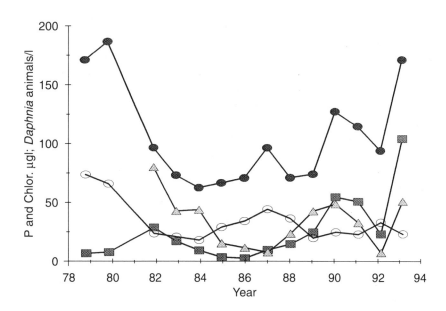

FIGURE 4.1 Changes in total phosphorus (●), soluble phosphorus (■), chlorophyll a (○) and *Daphnia* (△) in Cockshoot Broad, Norfolk
Source: Moss *et al.* (1996)

This resulted in a rise in chlorophyll levels, although total phosphorus did not change. Macrophyte populations also declined.

4   In 1989 the fish populations were removed, resulting in declines in chloro-phyll and increases in zooplankton and macrophytes. There was little change in total phosphorus or nitrogen. This stage continues today, with slight fluctuations.

The experience at Cockshoot Broad illustrates a number of important points. Removal of the primary source of nutrient input is vital. In this case sewage was fed via a river and this could have been prevented from entering the Broad. However, where this would not be possible, the installation of phosphorus stripping at sewage treatment works would be essential. Preventing further pollutant loading was also insufficient. Years worth of pollution had accumulated within the lake and this had to be actively removed by sediment dredging. Finally, while these treatments caused the desired changes in water chemistry and in some of the biota (e.g. a decline in phytoplankton and an increase in macrophytes), the role of fish populations remained important, allowing damaging phytoplankton blooms to occur. Thus additional biomanipulation was required in order to achieve the desired result.

CASE STUDY: USING BUFFER STRIPS TO MANAGE DIFFUSE
SOURCES OF AGRICULTURAL POLLUTION

The use of buffer strips to protect rivers from pollution is now widespread, and guidance on their use in England and Wales has been issued by the Environment Agency (EA, 1996b). A buffer strip may act in two ways. First, it puts a distance between the source of pollutants and the river. This may be sufficient to allow protection, for example from pesticide drift (see p. 147). Second, the strip may act to change the nature of pollutants before reaching the river.

A ploughed field may be a serious source of sediment loading to a river. A buffer strip of rough vegetation will trap much of the surface transfer of sediment during rainfall and prevent it from reaching the river. Obviously, the wider the buffer strip, the greater the effect. However, studies have found that most sediment is deposited within the first two metres of a strip and a 5 m strip will trap all but the finest particles.

Buffer strips may also affect dissolved pollutants. If the soils are water-logged, the anaerobic conditions will cause denitrification and so help to remove nitrogen. If there are serious problems with nitrogen or phosphorus, then harvesting of the vegetation of the buffer strip will help to remove some of those nutrients that have been absorbed in the plant material (provided that it is taken away and not left on the bank). The effectiveness of buffer strips for this purpose is highly variable and some strips may need to be up to 30 m wide to achieve significant results.

It is evident from the above description of buffer-strip function that they may perform different uses for different problems. It is important, therefore, for the pollution manager to have a good understanding of the nature of the pollutant problem. For example, should the buffer be waterlogged to improve nitrate removal, or well-drained to aid sediment removal? Some combination of characters may be possible, for example by balancing the mixture of grasses, shrubs and trees to vary the soil drainage characteristics. It is also important for the manager to identify the main hydrological pathways of pollutant movement. Water ingress from a field will not be uniform, but will depend on the topography. Where such ingress is concentrated it may be appropriate to have wider buffer strips or a strip with a different character.

Acidification

The production and deposition of acidic air pollution is described in detail in Chapter 2. One of the most significant effects (and one of the most publicised) of acid rain is its effects on freshwater. It is now known that the acidification has been occurring since the middle of the last century, as industrial emissions began to increase significantly (Battarbee, 1984), although it was the concern in Sweden in the mid-1970s over declines in fish populations and subsequent examination of changes in lake pH (Figure 4.2) that first raised it as an international issue.

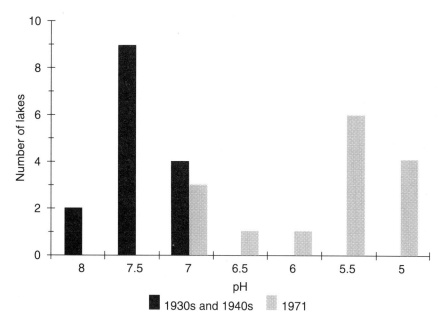

FIGURE 4.2 Changes in lake pH for a range of Swedish lakes over thirty to forty years
Source: Dickson (1975).

Very little rain falls directly on to lakes and rivers. It first passes over and through the soils and rocks of the surrounding catchments, and it is the nature of these catchments, combined with that of the rain, which determines the impacts on water bodies. If the soils and rocks contain significant buffering capacity (i.e. base cations such as calcium are easily weathered), then acid inputs are neutralised and water bodies are rarely affected. Such rocks include limestones, calcareous sandstones and ultrabasic igneous rocks. Acid igneous rocks (e.g. granite) have little buffering capacity and water bodies on these substrates are highly sensitive to acidification. Rainwater can reach water bodies in catchments through various routes. These include baseflow (when the water table is low, water percolates through bedrock), throughflow (through the soil horizons), overland flow (heavy rain resulting in water movement over the surface) and pipeflow (movement through cracks and channels). The first two of these has the greatest contact with and, therefore, the most effect on, the water chemistry of the rainfall.

Apart from lowering water pH, acid rain also affects other aspects of its chemistry. Beneficial base cations, such as calcium, may be depleted. Some aquatic species can be severely affected by this. Snails, for example, require a good calcium supply to build their shells. However, of most importance is the increase in other elements leached by the acid from the surrounding soils and, of these, aluminium has the greatest impact. Aluminium becomes particularly soluble below pH 5.0. It may stay as an inorganic cation or form hydroxy complexes. This latter form is especially toxic to fish and will result in high mortalities among their eggs and fry. This effect can be somewhat ameliorated in lakes with either high levels of organic matter or high calcium levels. However, with water of pH 4.5 and calcium at 50 μeq./l, Howells (1984) states that fish populations will decline when aluminium reaches 250 μeq./l. This has become all too common in a number of acid-sensitive areas of Europe and North America, e.g. southern Norway and Sweden, the Adirondacks in New York State, and in south-west Scotland.

Acidification affects all of the different classes of biota in a water body. Aquatic plants are affected. Some, like *Sphagnum* mosses, are actively promoted by the combination of low pH and low calcium. Indeed, *Sphagnum* actually produces hydrogen ions from its cells and so increases the acidification effects. It will overgrow other plants, and acid lakes will lose important species (Farmer, 1990). While *Sphagnum* bogs are rare and important, the soft-water systems that acidification may cause them to replace are even rarer. Other plant species may show a direct decline as pH levels fall.

Acidification of freshwaters has dramatic effects on the invertebrate populations. Some of the effects are indirect, i.e. changes in plant communities may affect those species which use them for food or shelter. Some impacts will be direct, by acting on invertebrate physiology. However, experiments tend to show that many invertebrates show less of a toxic response to aluminium levels than many fish. The relationship between invertebrate community assemblages and the acidity of lakes and streams is so strong that invertebrates are now used as biomonitors of stream chemistry (Petterson and Morrison, 1993). For example, circum-neutral streams (i.e. those between pH 6.0 and 7.5) are typically populated by mayflies and caddis-flies, while acid streams have lower diversity, with a typical group being the stoneflies.

Table 4.15 brings much of this information together in an examination of data from Welsh streams.

Of course other animals are dependent upon invertebrate and fish populations. There is some evidence that otter numbers, for example, are lower on acidified

TABLE 4.15 Fish and invertebrate status of Welsh streams, including the mortality of caged trout

Class	Fisheries status	Number of insect taxa	Trout mortality	Gill Al accumulation ratio
I	Abundant	70	0%	1:1
IIA	Abundant	61	0%	1:1
IIB	Absent	30	80–100 %	36:1
IIIA	Scarce	46	40%	7:1
IIIB	Absent	36	60–100%	57:1

Source: Ormerod (1991)

Note: Gill aluminium accumulation is measured as a ratio compared with the water in which the fish occur. Stream classes I, IIA and IIIA are in pasture catchments and classes IIB and IIIB are in coniferous forest catchments.

streams. However, the most extensive work has been on the insectivorous bird, the dipper. Dipper numbers are very closely correlated with stream acidity. The birds are highly sensitive to invertebrate numbers, especially mayflies and caddis-flies. Increasing acidity increases breeding territory size, reduces clutch size, brood size and chick growth rate (Ormerod, 1991). The effect of acidity not only affects the quantity of food, but its quality. Thus, in acid streams, food items contain less calcium, thus reducing the blood calcium levels in dippers and the thickness of the eggshells (which are, therefore, more liable to accidental breakage).

Using diatoms to reconstruct historic changes in lake acidity

Diatoms are unicellular algae, which construct hard silica cell walls that are unique to each species. It has been noted for a number of years that the particular species groups present in a water body are highly sensitive to water chemistry, including pH, alkalinity or aluminium. The hard silica cell walls do not readily decompose. Thus as the cells die and they deposit on to the bottom sediments of lakes, they accumulate to form a record of the species present at any given time. By taking cores of sediment from lakes it is possible to examine changes in diatom populations and to see how species abundance has changed. By linking these species changes to known preferences for water chemistry, it is possible to describe how the lake chemistry has also changed.

The sediment cores have to be dated in order to determine when changes may have occurred. This is most commonly done using radio-isotope changes to lead-210. This isotope has a half-life of 22.26 years, so levels of lead-210 allow accurate dating. Other components of the sediments can also be incorporated. For example, combustion of fossil fuels results in the production of small carbonaceous particles, and these remain in lake sediments. It is often noted that declines in lake pH do coincide with the appearance and increase in abundance of carbonaceous particles. Figure 4.3 shows an early example of pH reconstruction from the Round Loch of Glenhead in Scotland, showing changes in the abundance of different diatom species groups, lead-210 dates and pH.

Land management and acidification: liming

There are a number of different ways of managing freshwater acidification once the problem has arisen. The most common course of action is to attempt actively to reduce the acidity of surface waters by the application of lime.

Lime can be applied either directly to the water body that has been affected or to the catchment from which the water is drained. Lakes may be directly limed from the shore, by boat or even by helicopter. Rivers tend to be limed by a slow drip-feeding process from lime towers.

The country that has most enthusiastically embraced the use of liming has been Sweden. Each year, Sweden spreads more than 150,000 tonnes of limestone to

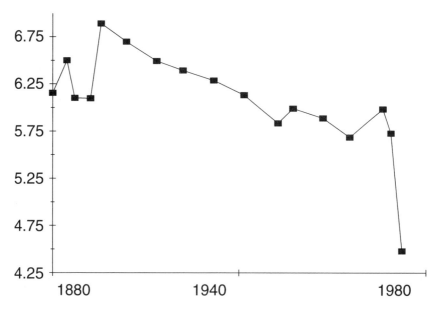

FIGURE 4.3 Changes in pH at Lake Gardsjon, Sweden, over the last 100 years, as determined by changes in diatom populations
Source: Renberg and Hedberg (1982)

PLATE 4.2 One way of examining the effects of acid rain on catchments is to roof the entire catchment and reintroduce 'clean' rain to observe recovery. This is a huge and expensive operation, but is a powerful demonstration of what acid rain is doing. This example is from Gardsjon in south-west Sweden.
Photograph: Andrew Farmer.

PLATE 4.3 Liming to reduce acidification of freshwaters is practised extensively in Sweden. Various methods are undertaken. This illustrates direct lake liming by boat.
Photograph: Andrew Farmer.

counter acidification. Much of this is spread on to lakes (from boats or helicopters), with additional spreading on to forest soils. The costs of the programme are over 100 million Swedish krona per year, and the aim is to extend the programme further. By 1990, over 85 per cent of the acidified lake area and nearly 6,000 of the 16,000 acidified lakes had been limed (Lovgren *et al.* 1993). The government generally pays 85 per cent of the costs of the programme, the rest being met by local angling clubs, local authorities, etc. While the programme began in southern Sweden, where severe acidification has occurred, it is now extending further north. The aim is to raise lake pH to pH 6.0, and there is a real danger than some northern, naturally acidic lakes will get caught up in the enthusiasm for liming and damage will be done to natural ecosystems. However, the use of lime has meant the re-creation and retention of many important fisheries and other freshwater uses as a stop-gap, while Sweden presses for reductions in acidic emissions from other European countries. Current commitments for emission reductions should already reduce the need for liming in some areas, e.g. Norway (Table 4.16).

There have been a number of problems with the direct liming of lakes and rivers. First, for example, many lakes have a short turnover period, so the lime can be flushed out too quickly. Thus reliming has to be done to retain the higher pH, and this can be very expensive. Second, it is not just low pH that is a problem for fish. The acid deposition in the surrounding catchment can leach metals such as

TABLE 4.16 Effects of decreasing the degree to which critical loads for freshwaters in Norway are exceeded, on the quantity of lime used to ameliorate acidification and its cost (1995 equivalent)

	1995	2010
Area exceeded (%)	29.7	10.8
Lime use (tonnes/year)	412,000	155,000
Cost (NKR millions)	335	124

Source: Henrikson *et al.* (1995)

aluminium into the lakes and cause locally toxic areas, e.g. where inflow streams occur. Finally, many fish use these inflow streams for spawning, and as eggs and fry are the most sensitive stages of the fish life-cycle to acidity, liming lakes alone may not be sufficient to maintain a viable population.

To solve these problems, catchment liming may be a reasonable alternative. With this method, the lime sits on the catchment and is slowly washed in. This reduces the frequency of application and helps to reduce local soil acidification as well. The use of lime can cause sudden loss of nitrate into the surface waters and this may be undesirable. However, catchment liming has proven successful at maintaining fish populations (e.g. Gee and Stoner, 1989). After early attempts at catchment

PLATE 4.4 Freshwaters may also be protected by liming forest soils, as here in Sweden. Photograph: Andrew Farmer.

PLATE 4.5 Remote areas, or those with easily disturbed soils, may be limed using a helicopter (as seen here in Sweden acquiring a new load of lime).
Photograph: Andrew Farmer.

liming, it was realised that the lime was being selectively washed from the *hydrological source areas* of the catchment, so the technique has been refined further to target these particular areas. These areas therefore receive higher rates of lime application, but the overall application rate to the catchment can be reduced.

The problem with catchment liming has been the nature of the hydrological source areas. For many of the acid upland catchments where the surface waters are acidifying, these areas are dominated by *Sphagnum* bogs. Such bogs are important for some rare plant species and are also important nesting sites for birds such as red grouse. *Sphagnum* is very sensitive to lime. It is killed by a combination of high pH and high calcium levels, and these are precisely the conditions that liming causes. The *Sphagnum* holds the bog together, and without it the peat may erode. Studies (e.g. MacKenzie *et al.*, 1990) have shown that three years after experimental liming, *Sphagnum*, liverworts and lichens are still adversely affected. There has also been a reduction in a number of invertebrate species, especially soil-dwelling species and this may, in turn, affect those small mammals which feed on them. The study concluded that 'no single optimum dose of lime would be compatible with the conservation of upland bog communities.'

It is also possible that liming can cause adverse impacts when applied more widely to soils. Recent studies, for example, have shown that widespread illness in elk in southwest Sweden may be due to the liming there. The liming leads to an increase in the availability of soil molybdenum, which can lead to a deficiency of copper within the bodies of the elk. This can cause a range of effects, from weight loss to problems with immune systems, blindness and heart attacks. Hundreds of elk have died so far.

Because of problems with hydrological sources areas in particular, it has been proposed that a series of issues have to be considered before catchment liming is undertaken (Farmer, 1992). These are:

1   To determine whether acidification has actually taken place. There have been proposals to lime acid lakes and ponds merely because they are acid. However, many of these lakes may be naturally acidic and so liming would not be justified.

2   If the lime application is merely intended to alter the status of the fishery of the lake, the original status should be assessed. For example, a study of the Adirondack lakes in the US showed that about half of the 'fishless' and acidified lakes had never supported good fish populations in the first place (Krester and Colquhoun, 1984). Liming these lakes may prove futile.

3   If the lake has to be limed, consideration should first be given to direct liming.

4   Where direct liming is not possible, the environmental costs of catchment liming should be weighed against the benefits. The areas that need to be limed will not all be of the same value, and assessments should be made of all locations that could be harmed prior to any application. That way, particularly sensitive and important areas can be avoided and the lime targeted on those of less importance.

5   Finally, of course, liming only treats the symptom and not the cause. Reductions in acid deposition are required for a long-term solution.

It may, of course, be necessary to use lime to protect species of purely nature-conservation importance. Sometimes direct liming of ponds can be used, e.g. as has been done to improve the breeding success of natterjack toads (Beebee *et al.*, 1990). Even catchment liming may be needed if lakes are used for important species and other measures are impractical.

A detailed study examining the potential for catchment liming has been undertaken in Wales, where eighty-three acid-sensitive catchments were surveyed for their vegetation interest (Ormerod and Buckton, 1994). The researchers found that while some of the upland catchments did contain important and internationally scarce mire communities, in some catchments only locally common vegetation types were present. It would also be possible to target lime applications to avoid important terrestrial vertebrates. It was, therefore, concluded that, with these data, liming of selected catchments could be achievable with minimal damage to terrestrial habitats of conservation importance.

Forest management and protection of freshwaters

While much of the publicity surrounding acid rain has focused on the effects that this pollutant is having on forest health, it is also the case that the presence of forests can exacerbate the effect of acid rain on surrounding freshwaters. The quantity of acid rain (whether in rain, mist or as gaseous pollutants) that is deposited is a function of the surface area of the receiving land. Forests increase the surface area tremendously and the canopies can efficiently scavenge pollutants. In effect, conifers can act as huge filters, removing pollutants in the surrounding air. The effect increases as trees grow and can be especially important in upland areas where mists are prolonged and extensive deposition can occur. Unfortunately, such areas also contain the most acid-sensitive freshwaters.

In the UK the Forestry Commission has produced guidelines on forestry plantations, which aim to reduce the impact that forests may have on surface water acidification (Forestry Commission, 1993). The guidelines are focused on new plantations (or replanting) and thus look to future impacts when current levels of acid deposition are reduced. Use is made of the UK's critical load maps (see Chapter 3) for freshwaters for acidity. These maps are based on the most sensitive water bodies in each 10 km square of the UK. If forests are proposed for such areas, detailed assessments (e.g. catchment-based modelling) should be undertaken to assess likely impacts on freshwaters. Simple changes in planting practice (e.g. maintaining wide buffer areas alongside streams with no trees or broad-leaved plants) can also help reduce acid impacts.

Plantation forestry may not only exacerbate the acidification of freshwaters. Significant problems can arise from the transfer of sediment during planting or harvesting, so that the levels of suspended solids can significantly affect the receiving waters. The Forests and Water Guidelines (Forestry Commission, 1993) include the following guidelines to reduce impacts:

◆   Only cultivate those parts of a site where it is necessary.
◆   Provide cross-drains to control run-off from cultivation channels.
◆   Align drains to provide an even gradient.
◆   Keep ends of drains away from watercourses.
◆   Construct silt traps at the ends of drains, and maintain them.
◆   Undertake work and maintenance at times of year when watercourses are less sensitive (e.g. when fish are not spawning).

Thermal pollution

Thermal pollution is the result of industrial activity that requires significant quantities of cooling water (Landford, 1990; Mason, 1991). This is largely from power generation. In Britain, the most affected river is the Trent, along which a large proportion of the coal-fired power stations are located. Inland nuclear power stations are also a significant source of heated water. The impact of a power station largely depends on whether the cooling water that is abstracted is recirculated and the heat reduced by the use of cooling towers. This is a common practice in Britain. Hot water discharged into a river or lake will mix slowly with the receiving waters, and a plume of water of elevated temperature is usually detectable. Commonly, the result of the discharge is to elevate temperatures by 3–5°C, although in some countries significantly higher elevations are found – of about 10°C.

The most obvious impact of rising temperatures is on the bacterial populations of freshwaters. Their growth is closely bound to ambient temperature and, given sufficient nutrients, will grow much more rapidly in warmer waters. This growth causes a significant increase in the oxygen demand on the water, reducing oxygen levels for invertebrates and fish. This effect is compounded by the fact that oxygen dissolves more slowly in warmer water. Some of this effect can be ameliorated where the thermal pollution is caused by power stations with cooling towers, and these towers will provide additional aeration prior to discharge to rivers.

Plants respond differently with rising water temperatures. In general, rising temperatures increase plant growth until a point is reached at which it declines. For example, growth of the flowering plant *Myriophyllum spicatum* decreases above 25°C, while that of *Potamogeton perfoliatus* is still increasing. Obviously, raising the temperature above a threshold for a given species will change the nature of the plant community. However, changes can also occur below this point, as the responsiveness of species will vary, so altering the competitive balance between them. Temperature has a stronger effect on respiration than photosynthesis. If these two metabolic activities are near to being in balance (e.g. in deep cool water where light is limiting), then increasing respiration may reduce the depth at which a species can colonise a lake or reservoir.

Fish may be particularly sensitive to thermal pollution, although direct death from heat is rare. Many species can detect temperature changes of only 0.05°C. High temperatures can cause heat stress in some species, although many fish can slowly adapt to temperature changes if introduced to them slowly. If the temperature rise is long term, then changes in fish behaviour will be seen, e.g. in the locations for feeding, and the location and timing of spawning. Experiments have been undertaken to determine the preferred temperatures for different species. For example, salmonids (salmon and trout) prefer temperatures around 13°C, while carp prefer 37°C (Varley, 1967). Thus different species will react to warm water in different ways. The low temperature preference of salmonids is important for migratory species, as salmon returning from the sea to inland spawning areas may find a warm stretch of river to be a barrier to continued migration.

A feature of ecosystem disruption due to thermal pollution is the survival of exotic species more commonly used to warmer waters. Such species may be accidentally or deliberately released. Examples include the flowering plant *Vallisneria spiralis*, the tropical tubificid worm *Branchiura sowerbyi* and the tropical fish *Poecilia reticulata*, which are all found in lowland English rivers where heated discharges occur. In unpolluted waters, their survival is prevented either because of their inability to survive the cold winters or because they require a critically high temperature to spawn. Thermal pollution may overcome these constraints.

Finally, it should be noted that thermal pollution can increase the toxicity of poisons in freshwater. A limiting factor for many toxins is the rate of uptake into the bloodstream and cells of fish and other organisms. Whether this uptake is passive or actively mediated by the organisms' metabolism, increasing temperatures will stimulate the rate of absorption.

Pesticides include herbicides, fungicides and insecticides. The general issue of pesticide pollution in the wider environment is beyond the scope of this book. However, it is worth highlighting some issues relating to pesticide pollution in freshwater in order to provide a more complete overview of pollution in this medium.

There are three main ways that pesticides can reach the aquatic environment: leakage during storage, spillage during transport and use, and directly from the use of the pesticides themselves (e.g. drifting from aerial spraying or leaching from field drainage). Obviously, the first two sources can be minimised by adoption of adequate housekeeping and safety procedures. However, the impact of applications themselves requires more careful consideration.

Pesticides are unlike most other forms of pollution. The production and use of the chemicals is designed to kill organisms in the agricultural environment. Thus, unlike most other toxic chemicals whose release is not meant to act to kill native organisms, pesticides are released for precisely that purpose. The task of managing their release, therefore, is either to find alternatives to their use or to ensure that impacts are focused on species targeted by the pesticides and that impacts on non-target species are minimised.

Species exhibit wide differences in their sensitivities to pesticides. Many aquatic species (both fish and invertebrates) are acutely sensitive to insecticides. However, toxicity tests on trout and a range of invertebrates have shown that *Daphnia pulex* is about 100 times more sensitive to the insecticide Carbaryl than trout, but about 100 times less sensitive to the insecticide Derris (Williams *et al.*, 1987). Insects are, for example, often more susceptible than crustaceans. Those that live on or near the surface and are air breathers, e.g. pond skaters, and are most exposed to aerial inputs or oil-based emulsions can, therefore, be most at risk. The impact of pesticides in freshwater also depends on a range of physical features, e.g. water depth and flow rate (increases in either of these will reduce pesticide impacts) and on the presence of organic matter, which may absorb some pesticides from the water column (although this may present a problem to some bottom feeders which ingest large quantities of such material). Concern has been expressed over the deposition of insecticides from aerial spraying (either from aircraft or tractors) on to sensitive water bodies that may hold important invertebrate or fish populations. Field experiments on four species of aquatic invertebrates have examined mortality at different distances from the application of insecticides (Pinder *et al.*, 1993). Figure 4.4 shows the results for Deltamethrin. The authors assume that invertebrate populations can generally withstand a mortality rate of about 10 per cent as being within normal natural fluctuations. They therefore suggest that a safe distance of 180 m for aerial spraying would be acceptable to protect those freshwater species studied.

While care in the application of pesticides (for example the use of safe distances) is essential, there are also a number of alternatives to pesticide use. Some of these involve traditional cultivation techniques, many of which have been reinvestigated by the current increase in interest in organic farming. Organic farming also employs

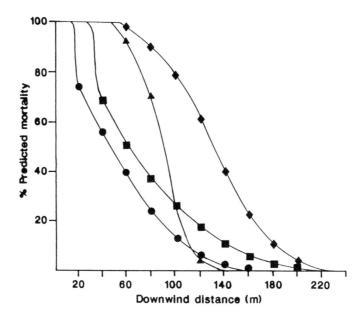

FIGURE 4.4 The predicted effect of distance from a target area on mortality among four aquatic invertebrates, following aerial spraying with the pesticide Deltamethrin: *Asellus aquaticus* (♦), *Gammarus pulex* (■), *Sigara dorsalis* (▲) and *Centroptilum pennulatum* (●)
Source: Pinder *et al.* (1993). Copyright is reproduced with the kind permission of English Nature.

newly developed methods for cultivation, particularly aimed at suppressing pests. Modern technologies can also be employed. For example, genetically modified crops may prove more resistant to pests, or modified insect viruses may help in pest control without using chemical pesticides. The release of genetically modified organisms involves a whole series of technical and ethical issues, which are also outwith the scope of this volume. However, it is important to understand that one beneficial effect of their use is to reduce the quantities of pesticides in the environment.

Environmental oestrogens

Recently there has been increasing concern over the potential effects of some pollutants in freshwater on the reproductive systems of animals and humans. These may act as hormones within the body, causing significant physiological changes, and are collectively referred to as 'environmental oestrogens' or 'endocrine disruptors'. Hormones are chemical messengers that travel within the body with the purpose of triggering particular responses. Usually, the body is especially sensitive to these chemicals and so it will react to them at very low concentrations. Thus for environmental pollutants, potential effects may occur at levels far lower than for many other pollutant impacts.

One source of environmental oestrogens is the artificial hormones manufactured for the contraceptive pill (e.g. ethinyl oestradiol) and discharged in sewage water. However, a range of other chemicals, e.g. pesticides, dioxins, and PCBs, may also have similar effects. However, recent research by the Environment Agency has shown that the most common problem from environmental oestrogens is the release of natural hormones from women (oestrone and 17B oestradiol) in sewage. On leaving the body these are inactive, but the sewage treatment process activates them. This area is currently under active research. Being female hormones, the greatest effect is in males, with reports of effects on lowered sperm counts and sperm quality, undescended testes, deformities of the penis and male breast cancer. For women, there are also reports of effects on breast cancer and on the cardiovascular system. Impacts are also seen in the natural environment, especially the masculinisation of female fish and effects on reptiles, molluscs and fish-eating birds.

It is still too early to be sure of the extent of this problem. While trends in human health characters (e.g. a lowering of the sperm count in males) are known, it is difficult to demonstrate that environmental oestrogens are to blame or what the overall social implications would be. Similarly, while effects on wildlife are likely to be due to the presence of these chemicals, the impacts on ecosystem functioning are unclear.

This is a developing issue and one that a pollution manager should be fully aware of. In particular, the issue of monitoring is especially difficult in this area. Experimental studies on the effects on wildlife indicate that changes may occur at concentrations that are so low that they are at the limit of detection for many types of instrumentation. It is possible, however, that some form of biomonitor may be developed to help with this in the future.

Groundwater

Groundwater is an important source of drinking water in many parts of the world, including the UK. It is also subject to pollution and measures need to be taken to protect it. A full examination of this topic is beyond the limits of this volume. However, a few key points can be made.

Groundwater is found only in particular rock types. Sandstones, chalk and limestone contain fissures, which become saturated with water (known as aquifers). This water can move from fissure to fissure through the rock. Water may enter these fissures at specific points, or through general seepage from overlying soils. The two most important groundwater-bearing geological strata in Britain are the Cretaceous chalk areas of the south-east and the Triassic sandstones of middle England.

Groundwater is a very significant resource. For example, the total volume in England and Wales is estimated at 16,000 million m^3, of which 10–15 per cent is accessible. This can be compared with the volume of surface waters of 40,000 million m^3.

Most of the pollutants which contaminate surface waters may reach ground-waters. Those which may affect human health are the most important, e.g. nitrate, pathogens, toxic organic compounds and heavy metals. Pollutants may actually be discharged directly into the groundwater, e.g. via a borehole or a well. Many others percolate through soils, e.g. from landfill sites or as nitrate from agricultural fertilisers. A serious problem in acid-sensitive areas may be the dissociation of aluminium and other metals from overlying soils that are subject to acid rain. In coastal areas, excessive abstraction from groundwaters may lead to intrusion of saline water into aquifers.

New survey work undertaken by the Environment Agency in England and Wales demonstrates that pollution threats to groundwater are very serious. The Agency found that 1,205 point sources of pollution were affecting groundwaters, and that 251 water abstractions were contaminated by these. The main cause (one-third) of these is leachate from landfill sites. However, when the different pollutants were examined for severity of impact, the biggest problem seemed to be that of organic solvents and pesticides.

Measures to protect groundwater are being undertaken. These include detailed monitoring of both quality and quantity, and an identification of pollutant sources and possible controls. The regulation of the agricultural use of nitrogen fertilisers described earlier is as much a protection of groundwater as surface waters, and illustrates this topic well.

The Environment Agency survey also examined whether remediation measures had been undertaken at these locations. The results (Table 4.17) show that attempts to deal with the problem only occur at a minority of sites. This is an area of fresh-water protection that deserves urgent attention.

TABLE 4.17   Types of remediation activity undertaken to control pollution to groundwaters from 1,205 point sources in England and Wales

Remediation activity	Percentage of sites with pollution control
None	56
Excavation	15
Capping/bunding	8
Pumping and treatment	8
Interception	4
Vapour extraction	2
Bioremediation	1
Other	6

Chapter summary

◆   This chapter examines the wide range of regulatory controls on freshwater
pollution. In the UK these are dominated by EU directives controlling the
release of dangerous substances and surface water quality. The Environment
Agency in England and Wales has an integrated approach to surface water
management, setting catchment-based plans taking account of all of the factors
affecting water quality and quantity.

◆   There are many sources of water pollution, including agricultural, industrial,
sewage, roads and domestic. These contain many pollutants, which can be
controlled in different ways:

    ◆   *Pathogens.* Viruses, bacteria and other parasites can pass into freshwaters
    if treatment does not occur, and may infect drinking or bathing water,
    resulting in a range of diseases, from mild diarrhoea to fatal attacks of
    meningitis or cholera. A range of disinfecting treatments are available,
    e.g. with chlorine or ozone.

    ◆   *Heavy metals.* Some of this contamination is natural, but much derives
    from industry, landfill sites, mining or sewage. Most effects on health are
    chronic, such as lead poisoning affecting mental development. Metals
    also affect aquatic life, although some plants, e.g. aquatic mosses, can
    absorb large quantities of metals, and thus these species can be used as
    biomonitors of this type of pollution.

    ◆   *Organic pollution.* This is derived from sources such as animal slurry or
    sewage, and results in a lowering of oxygen levels in the water, causing
    loss of fish and invertebrates. Some plants respond positively to organic
    pollution. A common effect is the growth of 'sewage fungus'.

    ◆   *Suspended solids.* These may be derived from many sources, e.g.
    agriculture or roads. The solids may smother aquatic life, changing
    animal communities. Only floating or emergent plants tend to tolerate
    such pollution. The pollution can be controlled by setting sediment traps
    alongside sources (e.g. drains, buffer strips, etc.) to allow sediments to
    settle out before reaching the water course.

    ◆   *Nitrate pollution.* Excessive nitrates in humans can affect the ability of the
    blood to carry oxygen. Nitrate pollution is primarily produced by the
    agricultural use of fertilisers. The EU now requires the identification of
    nitrate vulnerable zones, where controls on the use of nitrogen fertilisers
    are mandatory.

    ◆   *Eutrophication.* Excessive nutrient supply (nitrogen or phosphorus) will
    encourage algal blooms and the growth of some higher plants, choking all
    other aquatic life. Some blooms, e.g. blue-green algae, also produce toxic
    substances that may affect humans or their domestic animals. The main
    sources are agriculture and sewage, and these can be controlled through
    good management and secondary treatment. There is also a need to
    manage water bodies that have become eutrophic. This management may

include chemical treatments, removal of nutrient-rich sediments and control of inflow waters. Details are given of this type of management in the Norfolk Broads.

- ◆ *Acidification.* Acid rain (see Chapter 2) acidifies freshwaters. This not only lowers pH, but also mobilises toxic metals such as aluminium. Many invertebrates and fish are particularly sensitive. Some forestry practices may exacerbate the problem, and management controls on these are discussed. One management response to acidification is to lime soils or waters, and this has both advantages and disadvantages.
- ◆ *Thermal pollution.* Waste cooling-water from power stations may raise the temperature of some rivers. This affects plant and animal growth. In particular, in temperate countries such as the UK, the pollution may allow the survival of species from warmer climates that would otherwise not survive the winter or not experience temperatures that would allow them to breed.
- ◆ *Pesticides.* Agricultural use of pesticides is an important source of contamination. The role of safe distances situated close to watercourses in order to reduce impacts, is discussed.
- ◆ *Environmental oestrogens.* Small quantities of hormones from artificial sources, e.g. dioxins or the contraceptive pill, or natural hormones from women that are activated in the sewage treatment process, may result in sex changes in aquatic life. Concern has been expressed that sexual disfunction in some humans may be linked to exposure to these pollutants.
- ◆ *Groundwaters.* These are a major source of drinking water, and yet many have been polluted as they are 'out of sight'. Landfill sites are a major problem, with the most serious contaminants being pesticides and organic solvents.

Recommended reading

Dudley, N. (1990). *Nitrates – The Threat to Food and Water.* Green Imprint, London. This is a useful summary from an NGO standpoint of the problems of nitrates in drinking water.

Harper, D. (1992). *Eutrophication of Freshwaters.* Chapman and Hall, London. This book provides an excellent overview of the subject of eutrophication dealing with sources of pollution, the impacts that these may have and the range of management options that exist.

Haslam, S.M. (1990). *River Pollution: an Ecological Perspective.* Belhaven Press, London. This provides an overview of all aspects of freshwater pollution in so far as they affect the ecology of rivers.

Mason, C.F. (1991). *Biology of Freshwater Pollution.* Longman, London. This is a comprehensive account of the sources and impacts of freshwater pollution affecting both the natural environment and human health.

Smethurst, G. (1990). *Basic Water Treatment for Application Worldwide*. Thomas Telford, London. This is a good examination of the different techniques that can be applied to the treatment of wastewaters, especially from sewage. It describes in detail the different options, with their advantages and disadvantages.

Chapter 5

Pollution of the marine environment

◆ Introduction   156
◆ Types of marine pollution and their impacts   156
◆ Managing marine pollution   182
◆ Chapter summary   186
◆ Recommended reading   187

Introduction

The seas and oceans of the world have for centuries presented a picture of an endless landscape, a huge section of the environment that man was largely ignorant of, presenting threats to the survival to those working with it, and was impossible to control. For previous generations, it was almost inconceivable that waste disposed of in the seas could seriously affect them. In many ways the history of marine pollution is the pinnacle of the principle of 'dilute and disperse'. This principle is behind other pollution control measures such as uncontained landfill sites and the tall-stacks policies of the 1950s and 1960s for controlling air pollution. The idea is that toxic or nuisance substances lose their harmful qualities if dispersed into an environment where they can become highly diluted. In small quantities this is not unreasonable. However, as discharges increase in quantity, the principle breaks down. The sea was, for many years, viewed as an almost limitless location to dispose of and dilute waste from the land. However, we now realise that such discharges do pose serious threats to marine life and also to human health. The seas require protection, but many of the scientific and regulatory issues for a pollution manager are significantly more complicated than they are for the atmospheric or freshwater environment.

This chapter will examine different aspects of marine pollution, including aerial sources, garbage, heavy metals, oil, sewage and general eutrophication. It will conclude with an examination of the international and domestic regulatory framework that aims to protect the seas. Included in this is an examination of Direct Toxicity Assessment, a method of regulating industrial discharges equally applicable to marine and freshwater systems. The issue of radionuclide contamination of the seas will be addressed briefly in Chapter 6.

Types of marine pollution and their impacts

The marine environment is subject to a huge range of potential pollutants. The quantities and sources of pollutants vary widely in different locations. In inshore waters discharge of sewage or of oil-contaminated ballast water may predominate, while in mid-oceanic areas, airborne pollutants may predominate. Table 5.1 outlines the relative importance of different pollutant sources to the marine environment.

Pollution from the air

It seems surprising that the atmosphere is such a significant source of marine pollution. However, given the huge surface area of the sea and its close proximity to

TABLE 5.1 The relative quantities of different major sources of global marine pollution

Pollutant type	Percentage contribution	Example sources
Discharge from land	44	Agricultural waste, industrial discharges and sewage
Airborne pollutants	33	Industry and transport
Shipping	12	Oil spills, ballast water and other discharges
Dumping	10	Waste disposal (domestic and toxic) and shipping garbage
Offshore production	1	Sediments and oil

Source: GESAMP (1990)

many air pollutant sources, the potential for deposition is great. GESAMP (1990) identified the atmosphere as the most important source of a number of pollutants, including lead, cadmium, copper and zinc. Some of these may also be contained in domestic sewage or industrial discharges. However, in controlling these latter sources, it is obviously important to take account of the atmospheric inputs to ensure regulation would achieve its environmental objectives.

Garbage

There is an enormous amount of 'casual' waste entering the sea, and much of this can cause severe environmental problems. Fishing nets may become damaged and discarded, and yet continue to entrap fish and other marine life for many years. This is known as 'ghost fishing'. For example, it is estimated that about 30,000 northern fur seals die every year from entanglement with marine litter. The most persistent wastes are plastics, which are very resistant to biological or physical degradation. Large pieces of plastic sheeting may smother marine life and small objects are easily ingested by birds and marine mammals, leading to potential internal injuries.

Marine litter is defined by the National Academy of Sciences (1975) as:

> solid materials of human origin that are discarded at sea or reach the sea through waterways or domestic and industrial outfalls

In assessing litter that is found on beaches, for example, it is necessary to distinguish two further sources. The disposal of waste at sea, either industrial or domestic, is 'sea-borne' waste. That arising from the recreational use of beaches, etc., is 'land-borne' waste.

It is difficult to control litter disposal. Sea-borne sources are technically regulated through international conventions (see p. 163), but this is extremely difficult to enforce. While beach clean-ups are undertaken, these only have a limited value. The control of land-borne sources requires an educational approach, ensuring that people (whether visitors or residents) are more fully aware of the coastal environment and the consequences of their actions. A good way of raising the profile of marine litter is through participative monitoring programmes, which generate significant local publicity.

An example of litter monitoring is that of the Norwich Union Coastwatch UK. This has undertaken a survey every year from 1989, as part of a wider programme known as Coastwatch Europe. This provides comparative data across different countries into the major sources of litter and the threats that it might pose. The survey involves the use of volunteers who each assess standard sections of the UK coast and answer a common questionnaire about the nature of the coast and the litter that they find. For example, litter is divided into seven major categories and seventeen minor categories (e.g. medical waste, cans, plastic bottles, etc.). The main items found by the survey are plastic. This is understandable, as most plastic items may float well and will survive for many years without biodegrading.

It is also important to note that in some coastal waters and especially in the deeper sea the impact of shipwrecks can be a significant contribution to marine contamination. For example, between 1973 and 1995, over 29 million tonnes of shipping was lost world-wide (ILU, 1995). This compares with 21 million tonnes of Allied shipping alone lost during the Second World War. Many of the ships contained cargoes, which have toxic substances as well as general 'litter'.

Heavy metals

Discharges of wastewater from industry and from municipal sewers can lead to high concentrations of *heavy metals* in the waters and sediments of estuarine and other coastal waters (GESAMP, 1990). However, there are other sources. For example, it is estimated that the global burning of fossil fuels causes about 3,000 tonnes of mercury to be discharged annually (Ramade, 1979), and much of this ends up in the oceans. Table 5.2 indicates the main processes in the UK producing individual heavy metals. Some of these processes are large and represent individual industrial point sources of pollutants. Others are individually very small, but wide-scale domestic use may lead to significant quantities of metals being present in sewage. There are also very significant *diffuse* sources of metal pollution and other point sources, such as from landfills or from mining.

The effects that heavy metals may have on organisms vary greatly depending on the way that these organisms deal with them. For example, many bivalve molluscs are highly intolerant of most metals. However, many fish are able to deal with high levels of metals which, in low concentrations, are necessary for metabolism, e.g. zinc. Some species, however, seem to be able to accumulate large quantities of a wide range of metals, including non-metabolic toxins such as cadmium. In one study from

TABLE 5.2   Processes in the UK producing heavy-metal pollutants

Metal	Processes producing pollution
Arsenic	Wood preservative, glass production, alloys, medicines, semi-conductor manufacture, smelters
Cadmium	Pigment production, batteries, cement, agrochemicals, alloys, solders, photo-electric cells, electrodes, electroplating, photographic processes, deoxidisers
Chromium	Metal plating, iron and steel production, pigment production, textile colouring, lithography, photographic processes, glass and ceramic production, tanning
Copper	Metal plating, agriculture, alloys, wire, piping, dyeing, glass and ceramic production, vinyl chloride processes, wood preservative production, rayon, pigment production, anti-foulants
Lead	Batteries, leaded petrol, cables, solder, pigments, buildings, radiation shields, cement manufacture, fishing, shooting
Mercury	Batteries, agrochemicals, pharmaceuticals, mirrors, thermometers, barometers, chlor-alkali production, catalysts, dentistry
Nickel	Metal plating, iron and steel production
Tin (organic)	Anti-foulants, wood preservation, pesticides
Zinc	Alloys, electroplating, galvanising, rayon production, paper manufacture, fungicides, rubber, paint, ceramics, glass, reprographics, hygiene products

Source: NRA (1995)

the estuary of the River Thames in England, Rainbow (1987) found that 15 per cent of the body weight of some barnacles was zinc, sequestered away from metabolic activity.

Metals may enter organisms via a variety of routes. Those which are ingested are often readily absorbed. However, those dissolved in the water column may be less so, and will depend on the degree to which such metals may pass through membranes, e.g. those covering the gills. Cadmium, for example, may be actively taken up by fish by the ion-exchange pumps used for absorbing calcium. Tributyltin (see Case Study on p. 162) may be readily absorbed into fatty tissue due to the organic nature of the molecule.

Some metals can have serious consequences if they *bioaccumulate* in prey. Mercury accumulates in the feathers of birds and it is possible, therefore, to examine museum specimens to determine how concentrations have changed over time. For example, the fish-eating white-tailed eagle of Sweden had average feather mercury concentrations of about 6.6 ppm between 1880 and 1940. However, by the mid-1960s this had risen to 50 ppm (Clark, 1992). The decline of these sea birds in southern Scandinavia has, at least in part, been attributed to marine mercury poisoning.

Metals can have a wide range of toxicity symptoms. In animals they may cause *chronic effects* such as reduced growth and reproduction, damaged gills, etc., as well as the *acute effects* of severe toxicity. Metals may bind to proteins rendering certain enzymes inactive (including many repair enzymes) and they often interfere with *ion-exchange* processes across cell membranes. In plants heavy metals affect a range of metabolic processes, with electron transport during photosynthesis being especially sensitive.

There have been considerable changes to the discharge of metals into the marine environment in recent years, especially in response to international agreements, including EU legislation. For example, the black and grey lists of dangerous substances identified in the Dangerous Substances Directive (see details in Chapter 4) apply equally to marine as well as freshwater discharges. Table 5.3, for example, shows changes in the discharge of five heavy metals from England and Wales from all sources between 1985 and 1993. In all cases there has been a reduction, although this has been greatest for tin where special controls on tributyltin have been established (see Table 5.3).

TABLE 5.3   Loading of heavy metals (tonnes/year) to the sea from discharges from England and Wales in 1985 and 1993 and the percentage reduction achieved

Metal	1985	1993	% reduction
Cadmium	64	22	65
Copper	1,098	439	60
Lead	730	532	27
Tin	25	5	80
Zinc	3,340	2,360	21

Source: NRA (1995)

In order to manage heavy-metal pollution in the marine environment, it is necessary to understand the sources of the pollution, its distribution in the environment and how it might disperse. In the UK the most severe marine heavy-metal pollution is associated with estuaries, as these not only receive potentially contaminated river water from the land, but they are also the site of a large number of industrial activities that discharge directly into them. A case example is given for the Humber estuary.

It is worth examining some individual metals in more detail.

1   *Arsenic.* While arsenic is a highly toxic substance, it does not readily accumulate in the food chain of marine ecosystems, although elevated levels may be found in some shellfish.

2   *Cadmium.* Cadmium is an important metal pollutant that can be readily bioaccumulated by some marine organisms, including fish and invertebrates, as well as being found in elevated levels in some marine algae. High tissue concentrations in some shellfish have been known to result in illness in humans which have consumed them. Sediment concentrations may also be very high in areas where discharges occur, as cadmium is strongly absorbed on to the *cation-exchange sites* of the sediment particles.

3   *Chromium.* Chromium is bioaccumulated in some marine organisms, although this is less important than for some other metals.

4   *Copper.* Copper is an essential micronutrient, which is toxic in large doses. Indeed it is used in some paints used to prevent growth on ships' hulls, although it is not as effective an anti-foulant as tributyltin (see p. 162). Some molluscs (especially bivalves) are particularly prone to accumulate large quantities of copper, and some species are sensitive and are absent from estuaries where significant copper concentrations are maintained. In some long-term studies the evolution of strains of invertebrates more tolerant of copper pollution has been noted.

   Copper is toxic to humans. However, it is very unlikely that there is a threat to humans from eating seafood. A lethal dose of copper is about 100 mg. However, copper imparts a foul taste to seafood at very low concentrations, so inadvertent ingestion of this quantity of copper is remote.

5   *Lead.* Lead is a widespread contaminant in marine systems, due to its diffuse sources from atmospheric pollution derived from motor transport (see Chapter 3). However, it is not a soluble metal and is readily absorbed on to sediments. High water concentrations are, therefore, very unlikely and there is little evidence that lead is accumulated in the food chain of marine ecosystems, although some shellfish do show high tissue concentrations in contaminated areas.

6   *Mercury.* Mercury is a highly toxic metal, which, through marine contamination, has been known to affect a number of human populations. It is especially toxic if it occurs in an organic form (e.g. methyl mercury is thirty times more toxic than inorganic mercury). Organic mercury is also much more likely to persist within the biological components of an ecosystem, while inorganic forms may be removed to the sediments. Mercury is a cumulative poison that affects brain function (much like lead – see Chapter 2). The pollutant can arise from a number of sources, including pesticide use and a range of industrial activities. Sediments may also be a source of mercury if resuspended, as bacterial action may convert inorganic mercury to methyl mercury, thus magnifying the toxicity.

   Mercury bioaccumulates, and fish higher in the food chain can have high concentrations of the metal. If there is severe contamination, consumption of such fish can be fatal. Clark (1992) describes a case in the mid-1950s in Minamata, Japan, as a classic instance of mercury poisoning, whereby industrial contamination of fish resulted in 2,000 cases of poisoning, with forty-three fatalities and 700 permanent disabilities.

7   *Nickel.* Nickel is not very soluble and may rapidly be removed to the

sediments, but it can be found in the water column, in complexes with other metals and as organic molecules. It is not highly accumulated in the bodies of fish, though some shellfish may exhibit quite high concentrations.

8 *Zinc.* Zinc is the most mobile of the heavy metals described here as it readily forms soluble complexes with a range of organic and inorganic molecules. In very low concentrations it is required as a micronutrient in marine plants and animals. However, in high concentrations it is toxic.

Environmental quality standards for heavy metals

An important tool in controlling heavy-metal pollution is setting environmental quality standards (EQSs) for these substances, in order to assess the status and trends in water quality and the likely impacts of current and future proposed releases. In the UK some EQSs have a statutory basis in EU directives, while others have a non-statutory basis as recommended by the Department of the Environment. Current heavy-metal EQSs are given in Table 5.4.

TABLE 5.4 Environmental Quality Standards (EQSs) for heavy metals in the marine environment

Metal	EQS ($\mu g/l$)
Arsenic	25
Cadmium	2.5
Chromium	15
Copper	5
Lead	25
Mercury	0.3
Nickel	30
Tributyltin	0.002
Zinc	40

Note: All are based on annual average concentrations for the dissolved metal.

CASE STUDY: TRIBUTYLTIN

The toxic impact of most heavy metals to aquatic systems is certainly not the purpose of their release. Their discharge is a waste-management decision that may have toxic implications. An exception to this is tributyltin (TBT), which is released precisely because it is toxic. TBT is a pesticide. While pesticides are generally outside the scope of this volume, the localised effects of tin from TBT warrant its consideration alongside other metal pollutants.

Tributyltin oxide is added to paint used for the external hulls of ships. The toxicity of the compound acts to prevent the growth of organisms such as barnacles, and such paints are therefore referred to as 'anti-foulants'. TBT is very toxic and is designed to be so. Tin itself is a toxic metal. In the TBT the tin is combined with an organic molecule to allow its rapid uptake into the cells of aquatic organisms. It is also many times more lethal to most larvae of aquatic animals, as compared with the adults (a feature that helps to prevent the colonisation of ships' hulls). Acute lethal concentrations for larvae may be as low as 1.0 µg/l or even lower, and the no-effect level (*NOEL*) for the most sensitive molluscs is as little as 20 ng/l (Dobson and Cabridenc, 1990).

Some species of fish are also sensitive to TBT, though this varies considerably. TBT may affect the growth and reproduction of marine algae and higher plants, although this often requires higher concentrations than would cause significant effects on invertebrates.

TBT concentrations in sea-water have certainly been found to reach levels that have caused mortality among marine invertebrates. This has commonly occurred in the waters around boatyards, where ships are brought in for repainting.

TBT also has an interesting sublethal effect on some molluscs, e.g. dogwhelks. This may occur at concentrations as low as 1.5 ng/l. In these species, the female whelk produces a penis which grows over the oviduct, so preventing normal egg-laying. This condition is known as 'imposex'. Imposex can be readily identified and is a useful biomonitor for TBT pollution, as it may result from long-term exposure to very low concentrations that might not be detected adequately with intermittent chemical sampling. For those species that do not disperse readily (i.e. there is little recruitment from unaffected populations), extensive expression of imposex may lead to localised reduction in population numbers.

As TBT is extremely toxic, its use is now severely restricted in many countries. In the UK, for example, boatyards using TBT have now been included in this legislation as processes requiring authorisation under Integrated Pollution Control (see Chapter 3), as with any other industrial polluting activity.

CASE STUDY: HEAVY METALS IN THE HUMBER ESTUARY, ENGLAND

The Humber estuary is the largest in the UK; the rivers it receives drain one-fifth of the land surface of England, an area containing 11 million people. It is an important wildlife resource as well as having extensive fisheries and industrial activity. There are a large number of pollutant inputs to the estuary. Some of these are on the rivers that discharge into it and others emit effluents directly. An overall assessment of the impacts that these have has been made for the period 1980–1990 (NRA, 1993). Table 5.5 summarises the quantities of metal discharged to the Humber. In all cases, industrial sources are most important, although sewage outfalls contribute a significant proportion of some elements. Some of the sources of these metals can be identified as arising from a very few locations. For example, the non-ferrous metal

TABLE 5.5 Major loadings of metals to the Humber estuary from sewage and industrial discharges in 1990

Source	Metal loading (kg/day)						
	Arsenic	*Copper*	*Chromium*	*Cadmium*	*Nickel*	*Lead*	*Zinc*
Industry	375.4	69.9	601.8	3.0	41.5	63.1	1,422.3
Sewage	3.4	34.8	94.9	—	8.5	40.8	152.2
Total	378.8	104.7	696.7	3.0	50.0	103.9	1,574.5

Source: NRA (1993)

smelters on the north shore of the estuary are responsible for almost all of the arsenic and cadmium discharges, and the titanium dioxide processes on the south shore discharge most of the chromium.

It is important to assess the behaviour of these metals in the environment. Table 5.6 indicates the environmental concentrations of the metals discharged. In all cases, concentrations in the tidal rivers are greater than those in the open estuary. This is to be expected as these are closer to most of the effluent sources and less readily diluted by inputs from the North Sea. Even though there are large discharges, only concentrations of copper have been found to breach the environmental quality standard (EQS) for this pollutant.

Metals are not just discharged to the estuary and lost to the open sea, they may also accumulate in the sediments. Action by animals, storms or dredging may resuspend these and so affect water quality. Their presence in the sediments may also affect animals living there, with consequences for wildlife or fisheries. Currently

TABLE 5.6 Ranges in the average concentrations of heavy metals in the Humber tidal rivers, the Humber estuary and in estuarine sediments between 1980 and 1990

	Type of metal						
	Arsenic	*Copper*	*Chromium*	*Cadmium*	*Nickel*	*Lead*	*Zinc*
Tidal rivers (μg/l)	0.4–8.0	6.0–22.0	1.0–26.0	0.4–2.2	7.0–47.0	—	9.0–77.0
Estuarine water (μg/l)	2.0–12.0	5.0–22.0	0.5–2.0	0.0–0.3	3.0–15.0	—	9.0–14.0
Estuarine sediments (mg/kg dry weight)	21–51	37–76	42–81	0.7–1.7	28–49	59–117	185–333

Source: NRA (1993)

there are no EQSs for sediments, and the processes determining metal behaviour in sediments are still poorly understood. Table 5.6 indicates the sediment concentrations for the metals, and elevated concentrations are found around some of the larger industrial sources.

Any management of industrial discharges must take account of all current sources and the techniques that each source is using to reduce pollution. A process-by-process assessment is not sufficient. The overall environmental objectives for the estuary have to be set, e.g. for fisheries, and then the main obstacles to addressing this in terms of water quality have to be identified. These problems can then be related to individual discharges and management decisions taken to rectify them.

Toxic organic chemicals

There are a large number of toxic organic compounds that contaminate the marine environment. Many of these are pesticides used on the land, while others are the by-products of industrial or other uses. Table 5.7 provides a list of these compounds considered to be of highest importance in UK marine waters.

TABLE 5.7 A list of the major organic toxins considered to be of most importance in UK marine waters

Category	Substances
Organochlorine compounds	Chloroform, carbon tetrachloride, 1,2-dichloroethane, hexachlorobenzene, hexachlorobutadiene, polychlorinated biphenyls (PCBs), 1,2,3-trichlorobenzene, tetrachloroethylene, trichloroethane, trichloroethylene
Organochlorine pesticides	Aldrin, DDT, dieldrin, endrin, endosulfan, lindane, pentachlorophenyl
Organophosphorus pesticides	Anziphos-methyl, anziphos-ethyl, dichlorvos, fenitrothion, fenthion, malathion, parathion, parathion-methyl
Pesticides (other)	Atrazine, simazine, trifluralin

These toxic pollutants are known by a variety of collective names. One of these is especially apt – 'persistent organic toxins'. These compounds do not occur in any natural state and are not readily degraded by physical, chemical or biological processes in the sea. They do therefore persist in the environment for a very long time. Given their high toxicity, it is especially important, therefore, that their presence in the marine environment is monitored carefully and considerable effort is given to reducing contamination by such pollutants to a minimum.

It can be difficult to estimate the quantities of many persistent organic pollutants in the marine environment. They tend to occur in very low concentrations, sometimes near the limits of detection by monitoring equipment. Variations in river flows, for example, may make a large difference to the estimates of inputs, and errors in the estimates of loadings may occur. It was noted above that the total inputs of heavy metals in coastal environments around England and Wales are generally of the order of tens or hundreds of tonnes. Two of the more common organic pollutants are Lindane (an insecticide) and Simazine (a herbicide). It is estimated that the total loading of these pollutants in the seas around England and Wales in 1993 was 308 kg and 1,861 kg respectively. This is nearly two to three orders of magnitude lower than for heavy metals, and therefore illustrates the problems of detection and monitoring. It must be appreciated, however, that lower quantities compared with heavy metals do not necessarily mean a lower impact: pesticides are highly toxic as they are specifically released into the environment to target and kill organisms.

Some persistent organic compounds in the marine environment are derived from industrial sources and can be deposited from the atmosphere. These include polychlorinated biphenyls (PCBs) and dioxins. Some consideration has been given to these in Chapter 3. PCBs, for example, were manufactured and used around the world until their adverse environmental effects were understood. Production is now extremely restricted, although there are still problems for disposal of existing·waste containing PCBs. It is estimated that around two million tonnes of PCBs have been manufactured and that, so far, around one-third of this has been dispersed into the wider environment, including the seas. However, it is important to note that PCBs may persist for a very long time in the marine environment and can be transported great distances. There has been considerable concern expressed, for example, over the occurrence of PCBs in the tissue of organisms collected from Arctic and Antarctic waters, many thousands of miles from the sources of many of the pollutants (see Chapter 2). There are 209 different forms of PCBs identified and their environmental affects vary considerably, although impacts appear to be greatest on the juvenile or larval forms of species rather than the adults (Eisler and Belisle, 1996).

There is considerable concern about the potential carcinogenic effects of PCBs and dioxins on humans. In the marine environment such compounds may have more dramatic and immediate consequences. For example, Clark (1992) provides evidence that sea birds in the Irish Sea, and seals in the Baltic Sea, have shown direct mortality and breeding failure, resulting from PCB contamination. There is, however, little evidence of significant impacts on human populations from marine contamination.

Oil

Oil spills are probably the most emotive of marine pollution events. However, while a tanker wreck may result in extensive newspaper headlines, much of the oil in the world's seas comes from other smaller sources, such as tankers discharging ballast water from oil tanks used on return trips, leaking pipelines or engine oil disposed of down sewers. It is estimated that the annual oil input into the world's oceans is between 2 and 3 million tonnes. Of this, just under one-half comes from marine transportation, including accidents (Table 5.8). It is also important to note that dramatic accidents do not just happen to tankers. In 1977 30,000 tonnes of oil were lost from a platform in the Ekofisk field in the North Sea, although environmental damage was small.

However, the loss of 400,000 tonnes of oil from the Ixtoc site in the Gulf of Mexico caused extensive damage to marine life along the Texas coast. In extreme cases, oil may be deliberately released into marine systems for political reasons. Most notable of these was the deliberate destruction of the Kuwaiti oil wells during the Gulf War by Saddam Hussain and the consequent loss of oil to land and sea.

TABLE 5.8   The relative contribution of oil from different sources

Source	Percentage contribution
Industrial and urban run-off	37
Marine shipping operations	33
Tanker accidents	12
Atmosphere	9
Natural sources	7
Exploration and production	2

Source: ITOPF (1987)

The amount of oil lost from shipping is closely linked to the state of the world oil market. For example, GESAMP (1990) describes how the change in oil pricing by OPEC in the mid-1970s (which led to the oil crisis) caused a significant reduction (about 25 per cent) in oil moved by sea over the following ten years, and a reduction in the number of reported accidents involving tankers and in oil spills.

Oil spills can cause extensive environmental damage, with some ecosystems being more vulnerable than others. For example, a 1986 slick in Panama killed large areas of coastal coral reefs, which are slow-growing and slow to recover. The impact on this ecosystem may therefore be considerably greater, in the longer term, than that on many temperate systems. Similarly, oil trapped in complex mangrove ecosystems is very difficult to manage.

Oil has direct toxic effects on marine organisms. In high concentrations, rapid mortality may occur. However, sublethal effects are also seen, such as the reduced growth of marine plants, physiological changes in molluscs, reduced breeding success and community changes. It is estimated that a water concentration of aromatic hydrocarbons of 50 µg/l would be needed to cause effects on fish larvae. The natural background level in the oceans is about 1 µg/l, although this varies (e.g. that in the North Sea may be up to 3 µg/l).

PLATE 5.1 Tropical mangrove habitats are particularly sensitive to oil pollution. The oil is readily trapped among the tree roots, which project above the sediments to obtain oxygen. There are also sensitive animal communities, which can be easily damaged. The habitat is also particularly inaccessible, so clean-up operations to remove oil are very difficult. Photograph: Andrew Farmer.

Oil pollution has severe consequences for bird life. Once oil becomes incorporated into birds' feathers, the protective value of these feathers for insulation and buoyancy can be lost, removing the layer of air trapped beneath the feathers in most species. The birds are also unlikely to be able to fly. Many severely oil-covered birds will die rapidly, some drowning in the water. The first action of a bird becoming coated is to preen itself, trying to remove the oil. This is often fatal, as it results in ingestion of the oil and the toxic substances that it contains. Thus even a mildly contaminated bird may die, and it may be too late for those birds that have been caught in order to clean them. Huge numbers of birds may die. For example, the wreck of the Exxon Valdez in Alaska in 1989 caused over 30,000 bird deaths. Other sea animals may also suffer. The most notable of these were the sea otters in Alaska, where the Exxon Valdez caused over 1,000 deaths and loss of the species from a significant stretch of the coast. It is thought likely that in this case, their naturally high fecundity will lead to a reasonably rapid population recovery.

In small quantities (or when heavily diluted), oil can still have extensive effects on commercial activities. Small quantities can enter fish and shellfish, imparting an unpleasant flavour, a process known as 'tainting'. Depending on the degree of contamination, these flavours may range from being mildly unpalatable to highly unpleasant. Obviously, fish affected in this way cannot be sold. Being an organic compound, the oil more readily dissolves in fats within the animal tissue, and so any species containing high fat levels may be especially vulnerable.

It is important to note that there may be wider social and economic effects of oil spills. For example, the Sea Empress oil spill in Pembrokeshire, Wales occurred in an area that employs 6,000 people in the tourist industry (about 15 per cent of the workforce), with an estimated similar number indirectly benefiting from tourism. The potential impact of televised scenes of polluted beaches on tourism was greatly feared and considerable effort was given to cleaning up beaches and counter the adverse publicity.

As with any pollution event, the effect of oil pollution on human activities, e.g. food production or tourism, is only partly due to the physical and biological processes of the pollutant action. Just as important are the perceptions of populations (both locally and more widely). For example, when the Torrey Canyon ran aground off Brittany in 1967, fish sales in many parts of France were severely affected, even though fish from the area of the spill were not on sale. The concern over market impacts is so great that many fishermen are keen for severe restrictions following spills. Thus, after the Sea Empress spill in Milford Haven, in 1995, many fishermen were concerned that any early lifting of a ban on fishing (for which they could seek compensation from the tanker or harbour operators) could lead to some contaminated seafood reaching the international market, and thus have very long-term negative consequences for future sales.

In order to examine the different effects that a major oil spill might have, it is worth examining two case studies with varying impacts that received extensive publicity and generated extensive public concern.

CASE STUDY: ENVIRONMENTAL IMPACT OF THE 1993 WRECK
OF THE BRAER IN SHETLAND

The Braer ran aground at the southern end of the Shetland Isles in January 1993. Over a period of a few days it released 84,700 tonnes of Norwegian crude oil under severe weather conditions. The wave and wind action prevented a normal slick from being formed. The oil type was also more prone to dispersion than other varieties and soon much of the oil was deposited on to the sea-bed, and much of this in deeper water. The severe wave action also produced an interesting additional effect, in that a small quantity of the oil was deposited in aerosols on to the nearby land surface.

Shetland has a very well-documented environment, and so a committee was established to examine the ecological impacts of the spill, based on good baseline data (Ritchie and O'Sullivan, 1994). These authors concluded that:

> Overall, the impact of the oil spill on the environment and ecology of South Shetland has been minimal. Adverse impacts did occur but were both localised and limited. The resilience of ecosystems and species populations has been powerfully demonstrated, and provides confidence and reassurance for the future.

They found the following results:

1   For sea-birds the number immediately killed was much lower than initially feared in comparison with other spills. Sublethal effects were nearly absent in all species and a reduction in population size was only found for shags and black guillemots around the wreck area itself. However, in the following summer there was no reduction in the rate of successful breeding by those individuals that remained.

2   Only negligible effects could be detected on otters, seals and cetaceans.

3   Fish showed contamination for only a short period, and by April 1993 the temporary fishing ban was lifted. In particular, populations of the most ecologically important species, the sand eel, were found to be unaffected. This species forms the base of many food chains and is the staple diet of many species (e.g. puffins). It also dwells close to the sea-bed and could be more highly susceptible to deposited oil. Contamination in shellfish, however, lasted significantly longer.

4   Commercial fisheries in the area are also dependent upon salmon farming. Following the spill, these were contaminated and tainted. However, by July 1993 this effect had disappeared.

5   Benthic communities did show some changes, with an increase in oil-tolerant species.

6   Coastal communities showed remarkably few changes and little toxicity remains in the sediments.

7   Some effects were found on nearby land vegetation, but these were highly localised.

There is no doubt that the particular circumstances of the Braer wreck combined to cause a significant reduction in its impact. The rough seas quickly aided the dispersion of the oil, which, if it had formed a slick, could have posed a greater threat to birds and mammals. It also aided the dispersion into deeper water. The mid-winter timing also meant that many birds were absent and some *benthic* communities were more resilient. Given the very high concentrations of birdlife in the area, a similar quantity of oil forming a slick in mid-summer would have had very different environmental consequences.

CASE STUDY: ENVIRONMENTAL IMPACT OF THE 1996 WRECK OF THE *SEA EMPRESS* IN WALES

The most recent (and third-worst ever) oil spill in UK waters was that of the *Sea Empress* at Milford Haven in Wales. In February 1996 it released 72,000 tonnes of crude oil directly into an area containing some of the most important marine wildlife sites in England and Wales. An extensive slick was formed, and large quantities of oil were washed up on to the coastline. A number of clean-up operations were used, including physical removal of oil from beaches and the use of 445 tonnes of chemical dispersants at sea. However, consideration was given to the damage that the treatments themselves might cause. Chemical dispersants were not therefore used within 1 km of the shoreline, so that they would not build up to high concentrations in the shallow water and themselves cause a threat. Some sensitive shoreline habitats (e.g. salt marshes) were also not cleaned as the mechanical operation would have caused extensive structural damage. About one-third of the oil was lost to evaporation, and the combination of natural and chemical dispersants removed about one-half of the surface slick.

This volume was written only seven months following the spill, and so it is far too early to examine whether any long-term effects will result from it (e.g. the consequences of the sedimentation of the dispersed oil). However, the immediate effects of the spill have been extensively documented (SEEEC, 1996). As with south Shetland, the high wildlife value of the area meant that extensive baseline data was available, with which to judge the changes following the spill. These effects included the following:

1   Oil reaching the rocky shoreline caused extensive damage to intertidal life. Very high proportions of some populations of limpets, periwinkles, barnacles, algae and starfish have been lost.
2   Soft shores also were extensively damaged, with large quantities of dead bivalves, sea-urchins and starfish being washed up.
3   A fishing ban was immediately imposed. However, fin-fish showed little or no contamination, and by the end of May the ban on these had been lifted. Shellfish were more heavily contaminated and a progressive lifting of the ban took longer.
4   Large numbers of sea-birds were affected. By the end of May around 6,900 birds

171

had been recovered, about half of which were dead. About 5 per cent of the dead birds were washed up as far away as Ireland. The RSPCA handled around 3,100 birds for cleaning – releasing over 2,000. The vast majority of the birds were common scoter and guillemots. However, even though only a small number of red-throated divers were found, these may have represented a high proportion of the wintering population. As with other spills, the number of birds recovered would have been only a small proportion of the total affected. The effects of the spill were also reduced due to it occurring prior to the breeding season, and the return, therefore, of species such as the puffin.

5   There was no evidence of any mammal deaths, although some seals were oiled. There was also concern over one population of terrestrial mammals, i.e. greater horseshoe bats located in a sea cave. However, this did appear to be unaffected.

6   Shoreline vegetation was damaged. Plants were seen to be physically affected in rocky habitats, dunes and salt marshes.

Preventing oil spills: the wider context

Many of the actions needed to reduce the risk of major oil spills are beyond the ability of an individual pollution manager, depending on the way that international merchant shipping is regulated, etc. These issues were identified by the inquiry led by Lord Donaldson following the Braer wreck, which reported in May 1994 (DoT, 1994), making 103 different recommendations to prevent pollution from merchant shipping. Some of the most important for the UK included:

1   Many tankers are part of poorly maintained or inspected fleets. Following the 1993 Braer oil spill in Shetland, the Donaldson report showed that 600 ships in 1992 had been detained in British ports as being unsafe. Lord Donaldson high-lighted the problem. Some of this has arisen due to an influx of Eastern European vessels (so called 'klondikers') and of those from nations where registration requirements (e.g. infrequent safety inspections, poor training and low pay) are kept to an absolute minimum. These are the 'flags of convenience'. Thus Cyprus, Greece, Malta, Liberia and Panama have registered 34 per cent of the entire world merchant fleet.

2   There is a need for effective port controls. This could go some way to dealing with flags of convenience, by requiring higher standards from ships visiting the UK. However, many ships in UK waters do not visit British ports, so wider international agreement, e.g. within the EU, is necessary.

3   The crews of many ships are inadequately trained.

4   Better facilities need to be created in ports, to enable the disposal of waste, to reduce the likelihood of its discharge at sea.

5   Broader consideration needs to be given to the routing of ships around the coast. For example, some waters are designated as 'areas to be avoided' and these ought to be extended. Lord Donaldson recommended the identification of Marine Environmental High Risk Areas, where there is both a great deal of

sensitivity of the marine environment and high levels of shipping, and where entry of ships can be prohibited or restricted. However, as movement of shipping is governed by international law, the UK government cannot implement this unilaterally.

6   The ability of emergency services to respond should be increased, e.g. by a greater provision of salvage tugs around the UK coast. There should also be improved reporting by ships and more detailed tracking by coastal authorities. This is especially important in busy shipping lanes, such as the Dover Strait.

Preventing oil spills: land-based sources

While most oil spills occur at sea, there is also the potential for pollution from oil terminals, where huge quantities of oil are loaded and unloaded. There are a range of standard techniques that need to be established in an emergency procedures response. For example, there should be a system of shore-based booms, mobile booms, sufficient dispersants, pollution storage areas, etc., to contain and deal with any spill.

A major terminal ought also to have a monitoring strategy to establish baseline information on the environment and be able to assess the impact or routine operations or accidental discharges. One of the largest oil terminals in Europe is the Sullom Voe terminal in Shetland, which receives a large proportion of the oil extracted from the North Sea. In order to manage its environmental impact, the operators established the Shetland Oil Terminal Environmental Advisory Group (SOTEAG). SOTEAG is a mixture of interests from the oil company, the local authority, environmental bodies, etc., representing all relevant interests. It is able to advise on procedural issues, as well as devising a commonly agreed monitoring programme to assess potential environmental impacts.

Managing an oil spill

An oil spill has an appalling negative image on a television screen. The sight of waves moving slowly as a thick layer of black oil covers them, sends strong messages to the public. It is not surprising, therefore, that much of the efforts to control oil once it has spilled have been on removing this image. This means two sets of operations – trying to accumulate the oil for removal (e.g. from beaches or with bunds on the water), or attempting to disperse it using a variety of chemical treatments (National Research Council, 1989). This latter treatment may not be the best environmental option. For example, the effect of such dispersants used following the 1967 Torrey Canyon oil spill lasted for around ten years – longer than the oil contamination would have done. However, it did produce a 'cleaner' image more quickly. Indeed, the use of dispersants in this instance was initially thought to have had wider consequences when over 50 per cent of pilchard eggs were found to have been killed by the dispersants. However, this loss of fish fry did not appear to affect the catch rate once this age class reached adulthood.

CASE STUDY: NATURAL DEGRADATION PROCESSES OF OIL IN THE SEA

If left alone an oil slick will undergo a variety of degradation processes leading to the dissolution of the slick. These include:

1   *Spreading.* In the early stages of a spill, oil will spread rapidly over the water surface, depending on the oil viscosity.

2   *Evaporation.* The lighter oils (e.g. kerosene) within crude oil will evaporate into the atmosphere. This process is enhanced by higher temperatures and by wave action and wind. Most hydrocarbons with a boiling point of less than 200°C will evaporate within twenty-four hours in temperate parts of the world. In early stages of evaporation, the high atmospheric concentrations in the immediate vicinity may even pose risks of explosions.

3   *Dispersion.* Wave action will break up a slick and cause it to form oil droplets, which may remain in suspension, resurface or sediment to the sea-floor. The rate at which this happens will depend on the viscosity of the oil and the vigour of the wave action.

4   *Emulsification.* Over time oils physically absorb water to form an emulsion. These emulsions are often highly viscous and so may reduce the effectiveness of other degradative processes. High wind speeds increase the degree of emulsification, as they cause rapid mixing between the oil and water layers.

5   *Dissolution.* Some hydrocarbons, especially the lighter ones, are more soluble than others. However, as these are largely lost by evaporation, the separation of oil fractions by dissolution tends not to be an important process.

6   *Oxidation.* Exposed oils will react with oxygen in the presence of sunlight to form tars. This is a slow process and of minor importance in the overall degradation of a slick. However, it is responsible for some tar formations found on beaches.

7   *Sedimentation.* The heavier oils will sink in the water column, and some lighter oils will do so if they combine with particulate matter, such as that thrown up by wave action from the sea-bed.

8   *Biodegradation.* Some marine bacteria will use oil as a food source and cause its slow degradation. These degradation processes do require nutrients such as nitrogen and phosphorus to be present, as well as adequate oxygen. These conditions tend to exist as oils float on the water surface, but not once they have been incorporated into the sediments.

Sewage

This is the biggest pollution threat to most inshore waters and was made the highest marine pollution priority by the United Nations Group of Experts on the Scientific Aspects of Marine Pollution (GESAMP, 1990). Very few sewage sources around the world are treated before discharge to the sea. Even basic sewage treatment to remove large objects such as condoms or sanitary towels is often not undertaken, and their presence can make life unpleasant on bathing beaches. Secondary treatment to remove toxic substances or nutrients is even rarer. Sewage may either enter the sea via sewage outfalls extending from beaches, or from dumping from ships of sewage sludge produced in sewage treatment works. The types of pollutants entering the marine environment from sewage are the same as would enter the freshwater environment, and have been considered in Chapter 4. However, the relative importance of the impacts of the pollutants may be quite different in the marine environment.

One of the most serious risks to human health comes from bacterial contamination of bathing water. Whereas bacterial contamination of rivers can cause drinking water problems, such potable supplies can be treated before delivery. However, bathers are directly exposed to bacterial and other contaminants, and controlling pollution sources is the only management option (other than to prevent the use of the worst beaches). Bathers have reported a wide range of illnesses that are likely to have arisen from contaminated bathing water, ranging from ear infections through to highly dangerous viral meningitis and hepatitis. However, as with freshwater, the most common bacterial problem is *Escherichia coli*. Most environmental health monitoring consists of a measurement of *E. coli* populations as a surrogate for the overall level of contamination by sewage pathogens. Some sea users have formed pressure groups to campaign to combat the problem. One such group in the UK is 'Surfers Against Sewage'. They argue that surfers are at higher risk than other bathers. The action of a surf board is to generate a fine mist from the surface layers of the sea, and it is in these layers that many pollutants may be most concentrated. Pathogens are not only ingested by bathers. Many may pass into the human food chain by contaminating seafood. The most-common route for this is the contamination of shellfish. These molluscs are basically sedentary animals, which feed by extracting particles floating by in the surrounding water. The particles from sewage outfalls are readily trapped in the feeding apparatus, and pathogens contained in them can be passed to the shellfish. The problem is so serious in some areas that shellfish cannot be harvested at all. In many places, they have to be maintained in clean water for several weeks prior to sale, in order to reduce the contamination by pathogens. This treatment is known as 'depuration'.

As with freshwater systems, the input of sewage to marine areas adds large quantities of organic material. This can be broken down rapidly by bacterial action. However, such activity reduces the oxygen concentration in the water and can lead to death of fish and other organisms through oxygen starvation. This affect can be exacerbated further by the thermal structure of the sea. During the spring and summer the upper layers warm more rapidly, trapping a layer of cold water beneath. A *'thermocline'* is, therefore, said to have developed. Little exchange may take place between these layers until the winter cooling. Organic matter reaching these lower layers may be degraded more slowly by bacteria, as the lower temperatures reduce respiration rates. However, once these lower layers are devoid of oxygen, the lack of exchange with the upper ocean can cause long-term oxygen depletion. For example, in 1981 a thermocline developed alongside a phytoplankton bloom in the German Bight section of the North Sea. In some areas, all bottom-dwelling fish were killed, along with a range of molluscs and other invertebrates.

CASE STUDY: A SMALL BUT EFFECTIVE PRESSURE GROUP
SURFERS AGAINST SEWAGE

While most members of the general public express concern over the state of bathing waters around the coast, their interest in sea bathing is so limited that little political pressure is brought to bear. However, one group have taken up the challenge. This is Surfers Against Sewage (SAS), a small pressure group based in Cornwall. Surfers have to contend regularly with the unpleasant aspects of sewage discharges (e.g. waste floating in the sea). However, of greater concern is the small number of bathers (including surfers) who have succumbed to debilitating illnesses after swimming in the sea. These illnesses have resulted in lethargy, and a range of body disfunctions. However, it has been difficult to pinpoint the cause. If marine pollution is to blame it has not been possible, for example, to identify which pathogens in sewage or toxic substances in industrial discharges may have been responsible.

SAS was formed in 1990 and have mounted a series of high profile campaigns that have attracted attention in the UK and in the EU. The media have covered a wide range of stories, focusing on marine contamination that otherwise would not have attracted public notice. In a short space of time the UK government has recognised SAS as one of the more effective environmental campaigning organisations. While many of the regulatory and policy decisions required to protect bathing waters are still being developed, and the effectiveness of SAS is ultimately not yet determined, there is no doubt that such policy development will take place in a much broader public forum than would otherwise have been the case.

Nutrient addition

The run-off of nutrients (nitrogen and phosphorus) from agriculture or sewage treatment works can cause *eutrophication* problems in marine areas, just as in freshwater. This is particularly true where extensive areas of agricultural land may occur around shallow, enclosed seas, e.g. the Baltic Sea. It is quite common, for example, to see excessive growth of green algae, e.g. *Enteromorpha* spp. and *Ulva* spp., around the outfalls of untreated sewage works. It is also possible for eutrophication to lead to phytoplankton blooms. These may form dense populations within the water column or appear as masses on the water surface. An example of the latter is *Phaeocystis*, which can produce an unpleasant, brown accumulation on beaches. A huge bloom of this in the Adriatic in 1990 caused a widespread reduction in the tourists frequenting the resorts there.

Little work has been undertaken on agricultural marine eutrophication in tropical seas. However, there is currently considerable concern over the health of the tropical reef system in Florida Bay in the Gulf of Mexico. Here there has been extensive die-back of turtle grass, and the coral communities are suffering from an as-yet undetermined disease. A range of causes may be to blame, including changes in salinity. However, one likely explanation (suggested by increasing algal blooms) is run-off of agricultural fertilisers from the large, nitrogen-intensive sugar-beet farms of southern Florida. In other tropical waters, excess algal growth from nutrient addition has been found to kill coral by smothering the coral polyps.

Some international agreements have sought to take steps to reduce nutrient inputs to regional seas. For example, the Oslo and Paris Conventions have each (in 1992 and 1988 respectively) agreed action plans to reduce nutrient inputs to the north-east Atlantic, a number of areas of which are subject to eutrophication. The plans examine inputs from a wide range of sources and the effects of the control measures proposed by different states. Tables 5.9 and 5.10 outline the relative importance of the different sources, the relative contribution from different states and the reductions in nutrient inputs that would have been achieved between 1985 and 1995.

TABLE 5.9   Estimated total inputs to the marine area for 1995 covered by the Oslo and Paris Conventions (north-east Atlantic) from municipal treatment plants and agriculture of signatories to the conventions, and the effects of planned and executed pollution control (measures given as a percentage change from emissions in 1985)

Country	Municipal treatment plants				Agriculture			
	Phosphorus		Nitrogen		Phosphorus		Nitrogen	
	1995 (tonnes)	% reduction	1995 (tonnes)	% reduction	1995 (tonnes)	% reduction	1995 (tonnes)	% reduction
Belgium	4,800	52	28,200	12	<1500	> 35	<30,900	> 10
Denmark	700	63	2,700	73	530	5	50,000	15
France	NI	NI	NI	NI	21,000	17	180,000	10
Germany	9,900	74	148,500	30	13,500	21	270,000	17
Netherlands	3,700	72	31,500	32	6,400	0	139,000	23
Norway	510	46	9,960	19	160	36	8,450	29
Sweden	190	43	5,100	29	200	38	11,400	27
Switzerland	<1,000	> 57	17,000	6	NI	NI	NI	NI
UK	NI	NI	NI	NI	NI	NI	NI	NI

Source: OPC (1993)

Note: Data for Germany represent the area of the Federal Republic prior to unification. NI indicates no information is available.

CASE STUDY: DEADLY 'RED TIDES'

Sometimes algal blooms can appear red in colour, and so are often called 'red tides'. They are not particularly common in European waters, but are more prevalent along the coast of western North America. One effect of eutrophication is to alter the

TABLE 5.10 Estimated total inputs to the marine area for 1995 covered by the Oslo and Paris Conventions (north-east Atlantic) from industry and atmospheres of signatories to the conventions, and the effects of planned and executed pollution control (measures given as a percentage change from emissions in 1985)

Country	Industry				Atmosphere	
	Phosphorus		Nitrogen		Nitrogen	
	1995 (tonnes)	% reduction	1995 (tonnes)	% reduction	1995 (tonnes)	% reduction
Belgium	3,400	38	18,000	38	NI	NI
Denmark	120	96	1,200	63	150,000	27
France	NI	NI	NI	NI	NI	NI
Germany	4,500	29	40,500	40	1,050,000	25
Netherlands	6,800	50	8,500	47	274,000	27
Norway	60	50	1,500	70	NI	NI
Sweden	85	47	900	25	395,000	18
Switzerland	<30	>80	<1,000	>0	NI	NI
UK	NI	NI	NI	NI	860,000	−13

Source: OPC (1993)

Note: Data for Germany represent the area of the Federal Republic prior to unification. NI indicates that no information is available.

structure of marine phytoplankton communities. In nutrient-enriched waters, therefore, the relative proportion of diatoms decreases and the populations of dino-flagellates increase. It is the latter which form the 'red tides'. Some dinoflagellates may even cause the sea to phosphoresce at night. In very high concentrations the phytoplankton can physically interfere with fish, for example by clogging their gills, and occasionally they may also produce toxic substances. Such blooms do occur around the coasts of the North Sea, and have occasionally led to extensive fish kills and loss in invertebrate food sources. For example, a bloom of *Chrysochondromulina* was responsible for fish kills at fish farms in Norway and Sweden in 1988. In some instances, the toxins produced in blooms are not toxic to fish and can accumulate in them and be passed on to humans where effects may occur (McCallum, 1968). The effect is limited to a few dinoflagellates, e.g. *Gymnodinium*, and the condition produced is known as 'paralytic shellfish poisoning' (PSP), which includes a wide range of symptoms from nausea to death. PSP can also affect populations of natural species. For example, in 1968 82 per cent of the breeding shags in the Farne Islands in England were killed by PSP.

CASE STUDY: NUTRIENT INPUTS TO THE NORTH SEA ALONG THE SCOTTISH EAST COAST

Some concern has been expressed over the potential effects of nutrient addition to the North Sea and the effects that this might have to the waters of the east coast of Scotland. Lyons *et al.* (1993) have attempted to produce a nutrient budget for nitrogen and phosphorus to determine the relative importance of run-off from land via the major rivers.

The sources of inorganic nutrients were divided into the following:

♦ Major river inputs.
♦ Deposition from the atmosphere.
♦ Marine sources from along-shore currents (mostly arising from the west coast of Scotland).
♦ Marine sources from offshore currents bringing water from the North Sea.
♦ *Benthic*, i.e. mobilisation of nutrients from sediments.

The results are given in Table 5.11. It can be seen that the most important sources of nutrients are those brought in from the wider ocean. However, Lyons *et al.* (1993) conclude that riverine inputs (deriving pollutants from agricultural and sewage sources) may have important localised impacts. For example, the supply of agricultural nutrients into the estuary of the river Ythan, north of Aberdeen, has resulted in changes in *macrophyte* composition.

TABLE 5.11   Summary of the relative importance of different sources of inorganic nutrients to the Scottish North Sea Coastal Zone

Source	*Nitrate* (kt)	*Phosphate* (kt)
Rivers: Tweed	5	164
Rivers: Tay	5	147
Rivers: total	24	863
Atmosphere	5	0
Along-shore currents	60	20,000
Offshore currents	300	60,000
Benthic	35	23,000

Source: Lyons *et al.* (1993)

CASE STUDY: DOES MARINE EUTROPHICATION CAUSE ACID RAIN?

While deposition of atmospheric pollutants can lead to the contamination of the sea, and there are obvious links between pollution of marine and freshwaters, it is, perhaps, surprising that there is a potential for marine pollution to result in additional air pollution.

In Chapter 2, the formation of acid rain (as sulphuric acid) is described as the oxidation of sulphur dioxide from burning many fossil fuels. However, sulphuric acid may be formed by the oxidation of other sulphur-containing compounds. One of these is dimethyl sulphide (DMS). DMS is produced by the metabolic activity of a number of marine phytoplankton. It is estimated that, on a global scale, emissions of DMS may be of a similar order to anthropogenic emissions of sulphur dioxide (Bates *et al.*, 1992). The rate of emission of DMS depends upon the metabolic activity of the phytoplankton. Thus emissions in the North Atlantic are seasonal, being very low in winter when metabolic activity is reduced by the low temperatures (Leck *et al.*, 1990).

There is some debate over what contribution DMS makes to acid rain. Studies in areas of Western Europe such as Wales (e.g. McArdle and Liss, 1996) show that only a few per cent of the sulphur precipitated in acid rain is derived from DMS. The source of the sulphur deposited on to land can be identified by examining the relationship between two sulphur isotopes in the precipitation. Sea-water sulphate (which gives rise to DMS) has a higher ratio of sulphur-34 compared with sulphur-32 than the sulphur in oil or coal. Precipitation samples with isotope ratios close to those of sea-water, therefore, will be dominated by DMS.

While DMS production in the open ocean may not contribute large quantities of sulphur to acid deposition, it is possible that this is not so for all locations. In particular, there is concern for marine areas which have become *eutrophic* due to discharges from the land. Marine phytoplankton activity in such areas is greatly increased, with the formation of extensive blooms, especially in spring and early summer. This may lead to high levels of DMS production. In Europe, for example, the eutrophic waters in parts of the North Sea and Baltic Sea around Germany and Denmark are very close to the highly acid sensitive areas of southern Sweden and Norway. The precise nature of the relative contribution of DMS to acid deposition is still being investigated. However, it does illustrate the need for a holistic, integrated approach to examining pollution issues affecting land, freshwater, the seas and the air.

CASE STUDY: AN INTEGRATED ASSESSMENT OF EUTROPHICATION PROBLEMS AND CONTROL IN THE EUROPEAN UNION

It is clear that eutrophication is a major problem affecting all aspects of the environment, whether it be atmospheric deposition of nitrogen compounds, discharge from sewage treatment works into rivers or run-off from agriculture into the seas. Most attempts to regulate such pollution have focused on individual media, such as the air or the sea. However, many of the problems (e.g. excess use of nitrogen in agriculture) are common to all. A recent study (Van der Voet, 1996) has attempted to produce an overall examination of nitrogen flows within the EU and to assess the benefits that forthcoming controls may have on different eutrophication problems resulting from excess nitrogen availability.

The study used a method called substance flow analysis, which examines all the flows of a substance within a defined geographical area within economic and environmental sectors – in this case nitrogen within the boundaries of the EU (including the whole of the North Sea). Figure 5.1 shows the results for the environmental subcomponent of the analysis, showing the origin and movement of nitrogen through different environmental compartments. The analysis showed that over 90 per cent of all the nitrogen is derived from anthropogenic sources. Indeed the only significant other source was ingress to the North Sea from the wider Atlantic Ocean. It was difficult to identify economic sector origins for nitrogen inputs to the North Sea. However, agriculture was shown to cause 90 per cent of nitrogen inputs to groundwaters and 57 per cent of inputs to the atmosphere.

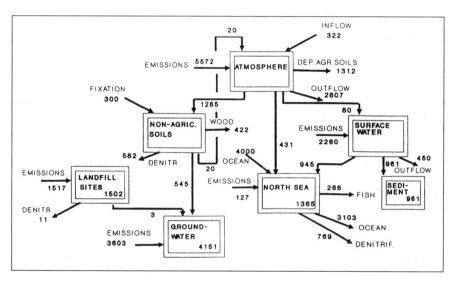

FIGURE 5.1 A diagram of the environment flows of nitrogen within the European Union in 1988, in kilotonnes nitrogen per year
Source: Van der Voet *et al.* (1996). Reproduced from the journal *Environmental Conservation*, with kind permission of Cambridge University Press.

A range of pollution reduction policies are currently in place in the EU, including reduction of nitrogen oxide emissions from industry and transport, improvements to sewage treatment and controls through the North Sea Conference and the Rhine States Conference. The study showed that if all the controls were fully implemented, they would result in a reduction by 40 per cent of the anthropogenic input of nitrogen to the North Sea and a 20 per cent reduction of anthropogenic nitrogen input to the atmosphere. However, for groundwater contamination little improvement would result. The activity of the largest producer of nitrogen pollution, agriculture, is dominated by the operation of the EU's Common Agricultural Policy. However, measures to reform the CAP to include sufficient measures for environmental protection are still in their early stages, even though it is becoming increasingly clear that such measures are long overdue.

Thermal pollution

The discharge of industrial effluent, especially cooling water from power stations, may cause localised increases in the temperature of coastal waters. For example, a typical 1,000 MW coal-fired power station would discharge about 20 million m^3 of cooling water at roughly 8–12°C above the ambient sea temperature.

As with freshwaters (see Chapter 4), the major impact of thermal pollution is to cause changes in species composition. Some cold-loving species may be lost. However, more noticeable is the survival of species that require warmer conditions, or changes in the behaviour of native species. For example, in the Firth of Clyde, Scotland, the warm waters from the Hunterston nuclear power station cause blooms of the copepod *Asellopsis intermedia* to occur two months earlier than in nearby waters.

The discharge of cooling water is regulated, alongside other emissions, under Integrated Pollution Control. The operation of a process would therefore require an environmental assessment of its potential impacts and these may be kept under review, e.g. with monitoring requirements, if impacts were thought to be possible.

Managing marine pollution

The management of marine pollution is often more difficult than for freshwaters or for air. It is especially difficult to control pollutants released at sea, e.g. litter and oil. Some of the issues concerned with managing the latter are discussed above. Those pollutants produced on the land (reaching the sea via rivers or the air) can be controlled more readily. However, even these may prove difficult, as many marine pollutants can be dominated by *diffuse* sources with less obvious means of regulation.

Chapters 3 and 4 discuss in detail the different ways in which emissions of pollutants to air and freshwater may be controlled. In these chapters emphasis has been given to the controls needed to give direct immediate protection to the air or freshwater environments. However, such controls are of course an option if the environmental impacts occur in the more distant marine environment. Similarly, technologies and techniques used to control freshwater pollution and described in Chapter 4 (e.g. buffer-strips for agricultural run-off or water treatment for industrial discharges) can equally be used for pollution sources around, for example, estuaries.

In assessing the potential impact of polluting processes on the marine environment, four key points were identified by GESAMP (1991):

1   Development must be undertaken in a way that avoids prejudicing environmental amenities for future generations (synonymous with sustainable development objectives – Chapter 1).

2   There must be avoidance of serious or irreversible harm to the environment.

3   There must be avoidance of measures that transfer damage from marine environments to other environments.

4   There is a need for concerted international action for environmental protection and preservation.

The regulatory environment, i.e. Integrated Pollution Control (IPC) in the UK and the new EU Directive on Integrated Pollution Prevention and Control (IPPC) as described in Chapter 3, applies just as much to the marine environment as to other media. Similarly, directives on water quality, e.g. the Urban Waste Water Treatment Directive or the Nitrate Directive, while delivering objectives relating to freshwaters, also have a key target of improving marine water quality. The Bathing Water Directive is more specifically linked to marine water quality. This does require specific action on sewage discharge direct to the sea, in order to control contaminants such as pathogens, although it also requires action to reduce discharge to rivers entering the sea. In implementing controls to protect the marine environment, action is often hampered by the fact that the marine environment is less clearly understood than that of the air or freshwater. This includes both the transport and behaviour of the pollutants themselves and the effects that they may have. In particular, this makes an integrated assessment of the relative importance of pollution to different media, as required under IPC or IPPC, more problematic. Some regulation other than of pollutants in sea water may, therefore, also be required. For example, there is some dispute over the adequacy of environmental quality standards for heavy-metal contamination as water concentrations. However, much of the regulatory effort on this in Europe is driven by the Shellfish Waters Directive, which sets limits on tissue concentrations in shellfish, which can be more clearly linked to the potential damage to human health.

Ultimately, there is a limit to what an individual country may achieve in terms of the management of pollution to its own coastal waters. It is certainly able to prevent the occurrence of high localised concentrations of pollutants. However, input of pollutants from neighbouring countries will occur, and international agreements are needed. Box 5.1 lists some of the large range of agreements that do exist. Some are under existing structures, such as the UN or EU, while others are at the behest of individual groups of nations. They may cover anything from strategic issues through to the specific management of one pollutant type.

BOX 5.1 International frameworks for protecting marine areas from pollution (and other damaging activities)

1 UN Convention on the Law of the Sea (UNCLOS). This became binding to participating states in November 1994. It outlines a framework of measures dealing with marine activities.

---

2 The International Convention for the Prevention of Pollution from Ships (1973). This is also known as the Marpol Convention. It was subsequently modified in 1978 and 1992. This is the most important international marine pollution agreement, as it has been signed by over eighty countries, covering over 90 per cent of the merchant fleet. It has specific sections covering topics such as oil, waste and rubbish, although each of these have to be ratified separately by each country.

3 The Convention on the Prevention of Marine Pollution by Dumping of Wastes and Other Matter (1972). This is also known as the London Dumping Convention. It covers a wide range of waste dumping issues, including that of radioactive waste.

4 UN regional agreements. Under the UN, countries in a particular area may sign regional agreements covering aspects of marine pollution prevention or protection. The UN coordinates these under its Regional Seas Programme. These include:

- ◆ Black Sea Action Plan (5 states)
- ◆ Caribbean Action Plan (Cartagena Convention) (25 states)
- ◆ East African Action Plan (Nairobi Convention) (8 states)
- ◆ East Asian Seas Action Plan (5 states)
- ◆ Kuwait Action Plan (8 states)
- ◆ Mediterranean Action Plan (The Barcelona Convention) (18 states)
- ◆ Red Sea and Gulf of Aden Action Plan (7 states)
- ◆ South Asian Seas Action Plan (5 states)
- ◆ South-east Pacific Action Plan (Noumea Convention) (5 states)
- ◆ South Pacific Regional Environment Programme (16 states)
- ◆ West and Central Africa Action Plan (21 states)

5 The Convention for the Protection of the Marine Environment of the North East Atlantic (1992). This was formerly the Oslo Convention and the Paris Convention. The UK is a signatory.

6 Helsinki Convention for the Protection of the Marine Environment of the Baltic Sea Area (1974).

7 The International North Sea Conference. This conference was first held in 1984 and has met three times since. On the second occasion (1984), a commitment was reached to control eutrophication, aiming to reduce anthropogenic inputs of nitrogen and phosphorus by 50 per cent between 1985 and 1995.

8 International Convention on Oil Pollution Preparedness, Response and Cooperation (1990).

9 Montreal Guidelines on Land-Based Pollution (1985).

10 European Union. Action includes Directives on Municipal Waste, Water Treatment, Bathing Water, Shellfish Directive and Nitrates Directive. A number of these are considered in more detail in Chapter 4.

BOX 5.2 Direct Toxicity Assessment

The traditional way to manage discharges to water (marine or freshwater) is to examine the range of substances to be discharged, and set individual limits on those chemicals according to environmental quality standards (EQSs). However, there are a number of problems with this approach, which mean that it is possible for operators to comply with current consent limits and still cause harm to the environment. Reasons for this include:

◆ The data for EQSs are limited – the vast majority of chemicals that are discharged (between 98 and 99 per cent) have no EQS, and others are poorly researched or even set below the ability of current monitoring techniques to detect such concentrations.
◆ Many discharges contain additional chemicals not necessarily identified in the effluent.
◆ Some chemicals may act together to produce an environmental impact. This *synergy* may be missed in individual chemical limits.

As a result, a number of countries in North America and Europe have adopted procedures for direct toxicity assessment (DTA). This uses a series of standard organisms to assess the toxicity of the effluent being discharged, irrespective of what it contains. Limits on toxicity can therefore be placed on the effluent, providing a clearer link to environmental damage. In 1987, the Organisation for Economic Cooperation and Development recommended that DTA be adopted in member countries, and in 1990 (NRA, 1990) a report to the National Rivers Authority in the UK stated that:

> For environmentally significant discharges of complex composition where not all important constituents can be individually identified and numerically limited, consents should specify a clearly defined toxicity limit, the appropriate form of toxicity test to be used, and the minimum frequency with which it should be applied.

As a result, in 1996, the Environment Agency, Scottish Environmental Protection Agency and Environment and Heritage Service in Northern Ireland proposed the adoption of DTA throughout the UK. Initially, at least, DTA would work alongside existing regulatory mechanisms so that individual discharge limits on specific chemicals and specific biological monitoring could still be required alongside DTA. The proposals highlight ten advantages that adoption of DTA can do to:

1 Provide clear, unequivocal target levels of toxicity that relate directly to the protection of aquatic life.

2 Provide a measure of aggregate toxicity that cannot be identified and licensed on a substance specific basis.
3 Be a measure of 'harm' and what is 'poisonous' as required by law.
4 Prioritise discharges for regulatory control.
5 Target monitoring resources by rationalising permissive chemical analysis [i.e. a non-statutory chemical sampling programme].
6 Prioritise expenditure for effluent treatment processes.
7 Provide advance warning of the risk of environmental damage (this is of particular value in any risk assessment strategy).
8 Improve the public image of industries shown to be releasing discharges of insignificant toxicity.
9 Provide information for general water quality assessments.
10 Aid in directing water quality improvement programmes.

The advent of DTA is therefore of enormous importance for the pollution manager dealing with discharges to either marine or freshwater systems, and close attention should be paid to the regulatory framework and toxicity procedures as they are developed and adopted by the environment agencies throughout the UK.

Chapter summary

◆ Marine pollution arises from many sources, and each may predominate in different areas. A range of problems are described. Litter is a serious problem for marine wildlife and aesthetic interests. It is also a pollution problem, with a relatively high public profile. Heavy metals are a particular problem in coastal waters, although their behaviour will vary, e.g. they can accumulate in the food chain or be removed to the sediments. A particularly toxic contaminant is tributyltin, a pesticide used as an anti-foulant, which can lead to loss of shellfish via a sublethal sexual disfunction known as Imposex. Organic toxins, such as pesticides and PCBs, occur throughout the marine environment, even in the most remote parts of the oceans. They are highly toxic and the release of these substances into the environment is now heavily controlled.

◆ Oil pollution has a very high public profile. Oil has a number of sources and effects. Many birds, for example, do not die from the covering of oil that they receive, but from the toxic substances that it contains, and which they ingest while attempting to preen themselves. Oil pollution can be controlled by adopting a range of management techniques to reduce risks. However, much of the action can only be taken through changes in international agreements or law. Those affecting the operation of shipping are included in the recommendations of Lord Donaldson's report on the Braer wreck in Shetland in 1993.

◆ Sewage discharge is very important, with resulting deoxygenation effects, which may kill fish. It also, along with agricultural pollution, causes eutrophication. One type of algal bloom caused by eutrophication is 'red tides'. These algae that form these tides may contain toxic substances, which can pass through fish to birdlife or humans, causing illness or death in a condition known as 'paralytic shellfish poisoning'.

◆ Many of the land-based controls on marine pollution are similar to those for air or freshwater pollution. For example, many EU directives regulate emissions to ensure overall protection of the aquatic environment. Industrial processes regulated under Integrated Pollution Control are examined for marine impacts alongside other environmental effects. A more recent development for regulatory decision-making is the development of Direct Toxicity Assessment (DTA). DTA will be applied in the UK to freshwater and marine discharges linking effluent impacts to standard toxicity tests, no matter what it contains, rather than specific emission limits to particular chemicals.

Recommended reading

Clark, R.B. (1992). *Marine Pollution*. Third edn. Clarendon Press, Oxford. This is an excellent review of marine pollution sources and impacts. It is a particular good source of information about organic pollution, oil, organic toxins, heavy metals, pesticides and radioactivity.

ITOPF (1990). *Response to Marine Oil Spills*. International Tanker Owners Pollution Federation Ltd., London. This is a useful handbook, leading the reader through the wide range of options available for oil-spill management, examining the advantages and disadvantages of each technique in turn.

NRA (1995). *Contaminants Entering the Sea*. National Rivers Authority Water Quality Series No. 24. NRA, Bristol. This report is a useful summary of basic data on the major sources of pollution entering the seas around England and Wales. It provides information on trends, individual sources, toxicity and regulation.

Managing pollution from radioactive substances

◆ Introduction   190
◆ What are radioactive substances?   190
◆ Measuring radiation   191
◆ Sources of man-made radiation   192
◆ The health effects of radiation   193
◆ The effect of radionuclides in the wider
  environment   195
◆ Radionuclides in the marine environment   203
◆ Regulation   205
◆ The peculiar case of radon gas   208
◆ Chapter summary   211
◆ Recommended reading   212

Introduction

The occurrence of radioactive pollution can be one of the most emotive subjects that a pollution manager may have to deal with. Releases around a nuclear power station may be examined with more scrutiny than significantly more toxic releases from other sources. Some of this fear is perfectly rational, arising from the fact that there is no safe level of radiation exposure and that radioactive substances may survive in the environment for thousands of years. However, there is also an important socio-political context for this reaction, arising from the atmosphere of the Cold War and the reaction to the potential use of nuclear weapons and thus a distrust of any form of nuclear technology.

This chapter will initially examine the nature of radioactive substances and how they are measured. It will detail the different sources of radiation, and how these affect human health and the wider environment, for both terrestrial and maritime systems. Two case studies are examined in detail. The first concerns the effects of long-term exposure to nuclear tests in the former Soviet Union, and the second the impact of the Chernobyl accident. The regulation of nuclear power in the UK, and how environmental assessments may be undertaken, will be described. Finally, a major source of exposure to radiation in Britain is from a natural source – radon gas. The occurrence, measurement, control and public reaction to this pollutant are described.

What are radioactive substances?

Radioactivity is produced by the spontaneous decay of the isotopes of some elements, whose nuclei are unstable. The radiation can take a number of different forms. In some cases it is as particles and in others it is electromagnetic. Five types of radiation may occur: alpha and beta particles, neutrons, gamma rays and X-rays. An alpha particle is large, consisting of two neutrons and two protons, whereas a beta particle is an electron. Gamma and X-rays have no mass.

The type of particles emitted is important in controlling exposure. An alpha particle, for example, cannot penetrate very far into matter as its size will inevitably result in collisions with surface molecules. A piece of paper is usually sufficient to prevent the movement of alpha particles. Given the small sizes of electrons, a thin sheet of metal would be necessary to stop beta emissions. However, the nature of gamma radiation means that up to 10 cm of lead is required to shield the environment from isotopes emitting this radiation product.

The type of radiation is not the only factor affecting the management of radioactive substances. The rate of emission is also extremely important. Some isotopes decay very slowly, others are so quick to decay that their existence is transitory in the extreme. The timing of the decay of any individual atom cannot be predicted. However, the probability of its decay is predictable and thus accurate estimates can be made for 'populations' of elements. This is described as the 'half-life', i.e. the time it takes for half of the atoms to decay. The half-life of sodium-24 is, for example, fifteen hours, but that for strontium-90 is twenty-eight years. An alpha source with a long half-life is therefore not particularly harmful. An example is carbon-14, which is used regularly in biological studies, and is assumed not to result in adverse effects. However, a short-lived gamma source (in sufficient quantity) is potentially extremely dangerous. As each isotope decays it produces new isotopes, generally of different elements. These in turn may also be unstable, although eventually the successive decay of these products will produce stable elements.

The length of the half-life of different isotopes is also an important factor in determining their abundance. Any short-lived isotope cannot be expected to be found naturally. While these may once have occurred, the age of the Earth is so great that their stocks will have decayed to undetectable levels. Only isotopes with long half-lives (e.g. uranium) remain. However, man-made reactions (e.g. in nuclear weapons or power stations) do generate these short half-life isotopes. Their occurrence in the environment is therefore a good marker for human contamination. For example, following the Chernobyl nuclear accident, large areas of Europe were contaminated, and this has usually been examined by assessing caesium-137 levels. This isotope has a half-life of twenty-seven years, so it does not occur naturally, but forms a useful means of tracing contamination both spatially and over time from man-made sources.

Measuring radiation

The task of the pollution manager is not helped by the fact that there is a range of different ways (and units) of measuring radiation and the effects that it has. As with many types of measurement, there is an older system and the current international system (SI). However, many published studies that are not particularly old do use the older nomenclature, and so it is important to identify the relationship between the different systems.

Measurements need to made of the quantities of radioactivity produced by a source, i.e. how quickly nuclear decay occurs or the rate at which the emission is occurring. The older unit for this was the curie (Ci), but this has now been replaced by the SI unit, the becquerel (Bq).

$$1 \text{ Bq} = 2.7 \times 10^{-11} \text{ Ci}$$

However, knowledge of rate of emission alone is not sufficient. For any assessment it is necessary to determine the dose that may be received by an individual. The simplest means of doing this was measured in rads, which are now replaced by the SI unit, the gray (Gy). However, dose is not just a factor of simple exposure, but is complicated by the energy of the different types of radiation, plus other factors. It also depends on how long a radionuclide will spend in the body. Thus a more biologically useful exposure measurement was devised, called the committed dose. This used to be termed the rem, but this has been replaced by the SI unit, the sievert (Sv).

$$1 \text{ Gy} = 100 \text{ rad}$$
$$1 \text{ Sv} = 100 \text{ rem}$$

Natural background levels of radiation are low and result from some input from cosmic sources (e.g. the Sun) and from naturally occurring isotopes. It is estimated that such exposure is globally about 1 mSv per year. Obviously, since the advent of nuclear technology there has been an increase in ambient levels of man-made radiation. It is estimated that the current dose due to previous nuclear tests (largely the above-ground tests that now no longer occur) is about 0.04 mSv, and that from nuclear power is only 0.003 mSv. However, very high doses have occurred in some locations, as will be described later.

Sources of man-made radiation

It is important to ensure that the potential risk from man-made radiation is examined in the correct perspective. The harm that a nuclear accident or weapons testing may do is enormous, and steps must be taken to reduce the risks. However, given the last five decades of use of nuclear technology, it is worth assessing how the radiation risks from all of the radiation so far released compare with those from natural sources. This is done in Table 6.1. It can be seen that previous above-ground tests have supplied by far the greatest exposure to radiation (much greater than Chernobyl). However, it is important to note that this is a global average and that local 'hot-spots' occur, where man-made radiation poses great risks.

TABLE 6.1 The percentage exposure to current global radiation from different sources, as compared with natural sources

Source	Exposure period/basis	Percentage compared to natural
Medical exposures	One year of current practice	24.7
Nuclear weapons tests	Cumulative to date	230
Nuclear power (normal)	Cumulative to date	2.7
Nuclear power (accidents)	Cumulative to date	5.5
Occupational exposure	One year of current practice	0.04
Natural	Global average	100

Source: Adapted from Gonzalez (1993)

Nuclear reactors

Commercial and military reactors both operate by the fission of uranium or plutonium atoms. The reaction creates a range of new elements or radionuclides. Some of these are heavier than uranium, others are lighter, and all have different properties to the original element. The containment of nuclear reactors is sufficient to prevent almost all releases to the wider environment. The most likely route of release is, however, through the cooling-water system, and under normal operating conditions small traces of contamination will occur.

The production of nuclear waste is a global problem, with global solutions being sought. Twenty-six countries now have commercial nuclear reactors, which number about 400. While the number of countries with military nuclear technology is much smaller, the amount of waste produced by the military at a global level is similar to that produced by commercial reactors. A typical 1,000 MW nuclear reactor produces about 25 tonnes of high-level waste per year, so the 400 commercial reactors that exist around the world produce about 10,000 t per year. However, only 10 µg of plutonium is sufficient to cause cancer in a human being, and it is estimated that one year's production of waste would be sufficient to cause hundreds of millions of cancers within three hundred years of production (Murray *et al.*, 1982). The only way to prevent such impacts is to isolate such waste from the human population, and thus there is a need for safe disposal facilities for high-level waste underground, which are being sought in a number of countries. The issue of nuclear-waste disposal is, however, dealt with in a separate volume in this series.

Nuclear installations also result in atmospheric discharges and discharges of liquid effluent. Much of this release is planned, although occasional accidents can result in small discharges in a few instances, through to the release of very large quantities of material, e.g. from the Chernobyl accident. It is usually accidental discharges from nuclear plants, rather than their routine operation, which release most radionuclides. For example, in 1996, the Dounreay reprocessing plant in northern Scotland released 7 GBq of radioactivity (mostly plutonium, americium and uranium) into the sea, probably due to a leak in the pipework which led to the contamination of the cooling water and the failure of other safety measures. This one event represented 13 per cent of the total average annual discharge from the plant.

The health effects of radiation

The effect of ionising radiation is to disrupt molecules within cells, thus causing chemical changes. It is possible to distinguish two distinct effects of radiation. At high doses, radiation causes burning, nausea and other rapidly produced symptoms. This results from the radiation causing extensive, immediate death of body cells. The effects are entirely predictable, i.e. if a person receives a certain dose, then particular symptoms will appear. Such effects are termed *deterministic*. However, at lower doses, radiation results in health problems such as cancer. It is possible to estimate the per-centage of individuals within a population that may be affected by a given dose of radiation. However, it is not possible to say whether any given individual exposed to that dose will become ill. These events are, therefore, termed *stochastic*.

The most important of these effects is the disruption to DNA, leading to the development of cancerous cell growth. It is important to note that there is no safe level for radionuclide exposure. Thus there is always a risk that the presence of ionising radiation may lead to detrimental damage to cells. Given that this type of radiation does occur naturally, albeit usually at low levels, the task of environmental managers is to reduce the exposure of populations to man-made radionuclides to as great a degree as possible, although absolute safety cannot be guaranteed as long as releases occur.

Routes of exposure

Radionuclides can enter the body via a number of routes, and each radionuclide may behave in different ways. Ionising radiation can enter cells either from external sources or internal sources (i.e. radionuclides that have entered the body, e.g. through ingestion). In general, less-penetrating radiation from internal sources is more damaging than more-penetrating radiation from external sources.

Some radionuclides are similar to common elements and compounds. Those elements with both stable and unstable isotopes will behave similarly, so that, for example, all iodine isotopes (radioactive and non-radioactive) will accumulate in the thyroid gland, so this is where the radioactive iodine will cause damage. Similarly, tritium (heavy water) will be readily ingested into the body and be transported and used by the body as other water is used. Thus this radionuclide will potentially reach all parts of the body. However, water is also readily lost from the body and so it is likely that a single ingested dose will be lost over a given period. Other radioactive elements will behave like their analogues within the periodic table. Thus strontium-90 will behave like calcium and so tends to be concentrated in bone tissue. However, caesium-137 will be utilised like potassium, which accumulates in body fluids. Some radioactive isotopes do not have stable analogues in any sense, but they may still show particular behaviour, e.g. plutonium is found to accumulate in bone.

While exchange is possible between radioactive isotopes and stable elements, many biological processes do tend to show a discrimination against the radioactive isotope. Plants, for example, will preferentially absorb calcium on to *ion exchange sites* in the roots when strontium is also present. Thus the ratio of calcium/strontium in the soil will be lower than that found in plant tissue. While it is easy to monitor soil contamination levels, it is important, therefore, to determine the barriers to transfer to the human food chain (or other sensitive receptors), before assessing the risks to health.

Much has been written about bioaccumulation, i.e. about pollutants that become more concentrated in tissues higher up the trophic levels. This has been particularly important, for example, in pesticide studies. There is also the potential for bioaccumulation of radioactive substances. However, their behaviour in this regard varies enormously and depends upon their position within body tissues. For example, there is little evidence of bioaccumulation of strontium. It is found in bone tissue, which does not tend to be absorbed when eaten by prey species. However, accumulation of caesium (within body fluids) occurs readily. For example, a fish may contain four times the concentration of caesium-137 compared with phytoplankton at the lower end of the food chain (Whicker and Schultz, 1982).

Some radionuclides, e.g. americium, are not readily absorbed in the gut. Thus if they are ingested, they may pass through the body with little effect. However, if dust containing the element is inhaled, it may enter the body. Americium is then transported to the bone marrow, where it accumulates and can cause cancers. Thus, although it is less likely to enter the body than tritium, once it has entered, it is more likely to lead to long-term damage.

The effect of radionuclides in the wider environment

There is a very wide degree of response to radioactive substances by different plant and animal species. Many of these studies have been undertaken with short-term acute exposure experiments, and data from these can indicate the dose needed to produce lethal effects (Table 6.2) Very few studies have, however, been undertaken in the field. There are obvious reasons for this. Releasing radioactive substances into the natural environment poses many long-term risks, especially as the aims would be to supply sufficient doses to cause significant adverse effects. Such experiments were, however, undertaken when regulation was not as strict as today. For example, Woodwell and Gannutz (1967) applied a large gamma radiation source to a mixed oak-pine forest in

TABLE 6.2   Ranges of acute lethal doses for different taxonomic groups

Taxonomic group	Acute dose range (Gy)
Viruses	200–10,000
Protozoans	100–4,000
Molluscs	100–1,000
Bacteria	50–10,000
Lichens and mosses	25–10,000
Insects	15–3,000
Crustaceans	15–200
Amphibians	8–50
Reptiles	8–40
Fish	7–60
Higher plants	6–900
Birds	5–15
Mammals	2–15

Source: Whicker and Schultz (1982)

North America. He noted that a significant change in the community structure of the forest was found with a dose of 1 Gy/day. Lichens proved the most resistant of species, while the pine trees were among the most sensitive. Much recent understanding of the impacts on the natural environment has come from examination of the results of unplanned releases, e.g. the Chernobyl accident or from nuclear tests. Results from these are provided in the case studies. However, they do have the disadvantage that many of the procedures for normal scientific experimentation (adequate controls, baseline data and monitoring) were not possible.

It is also important to recognise an important difference between plants and animals. Assuming that delivery of a radioactive substance is either a single event or continuous, it is most likely that a given plant will experience a fairly uniform exposure. Thus the dose received can be assessed and compared either with controlled experiments or to other 'real-life' situations. However, animals do move around. This means that their exposure will vary considerably over time and that the dose they receive will often be very difficult to calculate.

It has often been assumed that the protection of human health would, as a matter of course, lead to protection of the natural environment. This is based on the fact that humans are at the top of the food chain, and that whereas we are generally concerned with the protection of populations in the natural environment, for humans we are concerned with the protection of individuals (see Chapter 1). The loss of an occasional animal, for example, will not affect a population, but the death of an individual human is generally not politically or ethically acceptable. The assumption is that humans are at least as sensitive to ionising radiation as most other organisms in the environment. However, while this approach is unlikely to place wildlife *species* at risk, it may not adequately protect *populations*. The International Atomic Energy Authority has stated (IAEA, 1992):

> The human population group which is exposed may be separated geographically from a potentially exposed population or organisms, rare and endangered populations with very low fecundity may be present in the exposed area, and natural populations may be under ecological stress from a variety of natural or man-made pressures. Each of these situations should be assessed on a site specific basis and action taken to maintain the exposure rates within the limits appropriate for the organisms' protection.

In the UK, for example, a review was commissioned of the potential impact of radionuclides on wildlife (Kennedy *et al.*, 1990). This concluded that studies did not provide evidence of any deleterious effects on plants and animals in Britain, which could be attributed to environmental radioactivity. However, much of the existing monitoring of wildlife is based on the importance of wildlife to the human food chain (e.g. marine fish) and it was concluded that it is probable that some adverse effects do occur, but that these cannot readily be demonstrated under field conditions. It was therefore recommended that in very exposed situations (e.g. around reprocessing plants) wildlife should be monitored for levels of radioactivity, irrespective of whether it forms part of a food chain to humans.

CASE STUDY: LONG-TERM EFFECTS OF NUCLEAR TESTS

With the prospect of an international agreement to suspend all nuclear testing in the near future, it is important to remember that current testing procedures (i.e. underground tests) are still far safer than those undertaken in the early days of the nuclear era. Many of the studies on these impacts are now coming from the former Soviet Union, where data that were formerly kept secret are now being made public. For example, on the arctic island of Novaya Zemlya, ninety atmospheric tests were carried out between 1955 and 1962. This has resulted in increased contamination levels and the local indigenous peoples have elevated levels of lung, stomach and throat cancers.

One of the most used test areas in the world is that of Semipalatinsk in Kazakstan, a major test site for the former Soviet Union. Testing began there in 1949 and finished in 1989. Of the 470 tests, twenty-six were ground-level, eighty-seven in the air and 357 underground. The above-ground tests were all carried out in the early years of testing and so the effects of the contamination from the tests provide an interesting study of long-term radioactive contamination.

Sultanova and Plisak (1997) report that although the test site is itself enormous (18,500 km^2), radioactive precipitation has fallen over a much greater area of 304,000 km^2. This has included pasture lands important for local populations. However, the main impacts have occurred deep in the test region, away from populated areas. The general vegetation is one of steppe, dominated by grasses and sagebrush. The effects of the tests have been to produce a zonation of vegetation based on the degree of radiation exposure. Table 6.3 outlines the vegetation zones

TABLE 6.3   Vegetation zonation from the centre of the test area for above-ground tests at the Semipalatinsk nuclear test site

Radiation dose (Gy/h)	Vegetation present
108–82	*Artemisia frigida* as individuals. Ground structure affected by glass clinker formed by the blast
78–62.5	Sparse vegetation of *Stipa sareptana*, *Festuca valesiaca* and *Psatyrostachys juncea*
45–36	Vegetation occurring in sparse aggregations, with *Artemisia scoparia*, *Heteropappus altaicus* and *Kochia scoparia*
15–8	More diverse vegetation, with annuals and shrub species
2–0.8	Vegetation more similar to typical steppe vegetation
0.8–0.4	Vegetation indistinguishable from steppe vegetation

Source: Sultanova and Plisak (1997)

described in the study of the test site used for the initial above-ground tests. It is important to note that the workers found that, while species occurred in varying locations close to the test centre, there generally was a reduced occurrence of those that were reproducing. Thus the mere occurrence of a species does not indicate that impacts are not occurring.

Underground tests have resulted in a number of craters in the test area. One of these has formed a small lake and high radiation levels have been noted within it. However, even with exposure of 40–23 Gy/h, the vegetation of the lake edge was dominated by the reeds *Phragmites australis* and *Typha angustifolia* and these were fruiting. These plants seemed therefore to be more tolerant than many terrestrial species.

CASE STUDY: THE CHERNOBYL NUCLEAR ACCIDENT

On 26 April 1986 the worst nuclear accident in history occurred at Chernobyl. Chernobyl at that time was in the Soviet Union, and today lies just inside the Ukraine on the border with Belarus. The accident resulted in two massive explosions, blowing the 1,000 tonne plate off the reactor and the roof off the building. The graphite blocks used to regulate the reaction fell into the reactor, and a large fire raged, sending smoke high into the atmosphere. It was not until the 6th of May that the reactor was stopped. During this period it is estimated that up to 8 per cent of the one billion curies contained in the reactor were released to the atmosphere.

The pollution had a devastating impact on surrounding communities. Within the first few days of the disaster, 115,000 people were evacuated from a 30 km zone around the reactor. Since then it is estimated that about 28,000 km² of land in the Ukraine, Belarus and Russia, containing over 2,000 towns and villages, and 850,000 residents have been officially declared as contaminated (caesium-137 level in excess of 185 kBq/m²). Some 144,000 ha of agricultural land and 492,000 ha of forestry were removed from commercial exploitation.

The distribution of radioactivity from the accident was able to be detected in countries throughout the Northern hemisphere. Figure 6.1 shows the relative dose that a range of countries received. Much of the more distant deposition from Chernobyl was of light isotopes. However, this deposition was very patchy, depending on local weather patterns and the nature of the receiving vegetation. This provides added difficulties in assessing ecological impacts.

Impacts on human health

Managing the effects of the Chernobyl accident on human health

Savchenko (1995) describes a series of measures that were taken to manage the immediate effects of the accident and to reduce longer-term impacts. These measures were only partly successful. They included:

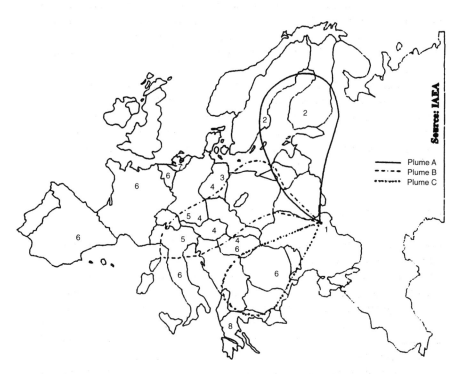

FIGURE 6.1 The dispersion of the plume and reported arrival times of detectable activity
in the air, following the Chernobyl accident in 1986
Source: Savchenko (1995). Reproduced with kind permission of UNESCO.
Note: Plumes A, B and C correspond with air-mass movements originating from Chernobyl on
26 April, 27/28 April and 29/30 April respectively. The numbers 1 to 8 indicate initial arrival
times: 1, 26 April; 2, 27 April; 3, 28 April; 4, 29 April; 5, 30 April; 6, 1 May; 7, 2 May; 8, 3 May.

- ◆ *Thyroid blocking.* The most pressing immediate problem was the intake of the
  short-lived isotope iodine-131, which accumulates in the thyroid, resulting
  in disfunctions of that gland. To counter this, 5 million people were given
  large doses of non-radioactive iodine to reduce the potential for iodine-131
  accumulation.
- ◆ *Food supply.* Uncontaminated food supplies were brought into the area to
  replace those affected by the accident.
- ◆ *Water supply.* New deep artesian wells were dug to obtain uncontaminated
  water.
- ◆ *Relocation.* A programme of evacuation and relocation took place. An
  exclusion zone was identified around the plant, and around this was established,
  first a compulsory relocation zone and then a voluntary relocation zone.
  Those remaining within the voluntary zones were subject to regular health
  assessments. By 1992, 163,000 people had been evacuated.

♦ Decontamination. A number of areas around the reactor site have been subject to decontamination procedures to reduce population exposure to radionuclides.

Genetic damage

While immediate changes to ecosystems and damage to human health can result from excessive discharges of radioactive substances, the more insidious effects of long-term damage to the genetic structure of those populations is less understood. Recent studies (Dubrova *et al.*, 1996) have examined changes in human populations around Chernobyl, demonstrating effects nearly ten years after the accident.

Following earlier studies reporting increased thyroid carcinoma in children around Chernobyl, elevated frequencies of chromosomal abnormalities in the general population and high rates of congenital deformities in newborns, Dubrova *et al.* (1996) examined mutation rates in children in the Mogilev district of Belarus near Chernobyl, who were born after the accident occurred, and compared these with a control population in Britain. The Belarus population were exposed to high levels of iodine-131 immediately after the accident and continue to be exposed to the more stable caesium-137. The rates of mutation were found to be twice as high in the Mogilev population, and a strong correlation was found with surface levels of caesium-137 contamination. Rates of mutation were also found to be higher than would have been predicted previously, given the exposure of some portions of the population to ionising radiation. If continuing exposure to existing contamination is producing these high human mutation rates, then the effects of the Chernobyl accident will pose a significant threat to these populations for many years to come.

Effects on flora and fauna

It is difficult to assess the precise effects of the ionising radiation from Chernobyl on the surrounding flora and fauna. The most dramatic effect was that the radiation killed many of the pine trees in the area. Around the Chernobyl site, a series of zones could be detected, which show progressive impacts on forest systems, dominated by scots pine (*Pinus sylvestris*). The areas given below are the full geographical extent of the zones, although only a proportion of them were covered in forest. These are outlined in Table 6.4.

However, such a change in the ecosystem structure, together with the loss of human inhabitants (and their animals) would, in itself, have tremendous impacts on the native species. Following the accident, for example, the numbers of game species increased markedly, taking advantage of the lack of human pressure. Some studies found no obvious effects on the populations of some species, e.g. soil microfauna, many plants and fish. However, longer-term examination of mutation rates in these species is, of course, not yet known.

Baker *et al.* (1996) found much higher than expected rates of mitochondrial mutation in voles close to the Chernobyl reactor. In fact, the mutation rate was about

TABLE 6.4   The relative impact of Chernobyl radiation on forests immediately around the accident site

Zone	Dose (Gy)	Area (ha)	Impact on forest
1	80–100	4,400	Complete mortality of mature pine trees
2	10–20	12,500	Majority of mature pine trees showed apical mortality, and many young trees showed complete mortality
3	4–5	11,900	Needle drop, causing severe impacts on growth
4	1–1.5	Rest of restricted zone	Changes to growth and reproductive processes

two orders of magnitude greater than predicted, and some individual animals were found to be highly contaminated with radionuclides. Part of this may be explained by the fact that, close to the reactor, radioactive contamination was not the only consequence of the accident, but there was also considerable pollution by other contaminants such as heavy metals.

The effects of high mutation rates will vary, depending on the type of organism in which it is occurring. Even with high levels of mutation and contamination, the voles around Chernobyl are thriving. The same can be said for a number of other species. A number of conditions are favourable to these species (Hillis, 1996). There is reduced competition for food, reduced predation and abundant sources of food. The mutation rate almost certainly reduces individual life expectancy, and possibly fecundity. However, reproductive rates for many species are high enough to maintain the populations (which may also be supplemented by outside migration). Thus a short-lived species with a high reproductive rate may survive the pollution pressures placed on it by an extreme event like Chernobyl. However, slower-growing species, with lower reproductive rates, are likely to be at greater risk. The long-term effects of Chernobyl acting on the genetic structure of the natural environment are not easy to predict.

Wider effects of Chernobyl

The Chernobyl accident proved a tragedy for the region in which it was situated. However, the widespread contamination also had economic effects across Europe. Aquatic contamination was found in a number of areas. For example, elevated concentrations of radionuclides were found in the Baltic Sea, the Danube and

Scandinavian rivers, and the contaminants were passed onto fish. However, only in a few isolated cases did tissue levels rise sufficiently to affect human consumption. The same cannot be said for terrestrial environments where radionuclide deposition posed significant threats to human health. Two cases highlight this – the effects on reindeer in Scandinavia and on upland livestock farming in Britain.

The effects on reindeer are a good example of biological processes enhancing exposure to radioactive substances. The main part of the reindeer diet consists of lichens. The large branched lichens in northern Scandinavia form dense layers, which, while being nutritionally poor, are available throughout the winter. Lichens also have another property, in that their cell walls have a high *cation exchange capacity*. Thus radioactive cations such as caesium or strontium will readily be absorbed on to the lichen tissue, producing concentrations of over 100,000 Bq/kg (fresh weight). As a result, the reindeer that eat lichens may ingest very large quantities of these isotopes. Before the Chernobyl accident, the Swedish standard for radioactive contamination of meat for human consumption was 300 Bq/kg. Following the accident, tests showed that 75 per cent of the reindeer meat produced exceeded this value and had to be discarded. In 1987 Sweden lowered the standard, but still significant quantities of the meat produced were unfit for humans. Interestingly, young animals

PLATE 6.1 Arctic lichen communities such as this one are able to accumulate large quantities of deposited radionuclides due to the large cation exchange capacity of their cell walls. They are the staple food of animals such as reindeer, and so the animal's meat may become contaminated and unfit for human consumption.
Photograph: Andrew Farmer.

were found to have higher concentrations than their mothers, suggesting some selective transfer within the womb and, as a result, the survival rates of the calves declined. However, the real threat was not to some individual animals or to the supply of a desired food item in Sweden. The inability of the reindeer herders to sell the meat threatened their livelihoods. Unfortunately, this is a traditional activity of the Laplanders, a small native racial minority in northern Scandinavia. The possibility of a significant threat to a particular culture was not foreseen, and significant efforts had to be undertaken to protect this cultural heritage. It should also be noted that other important animals such as the moose have caesium-137 levels in their meat above that fit for human consumption (Palo and Wallin, 1996).

Contamination of wildlife and crops was found throughout Europe following Chernobyl, e.g song thrushes in Spain, foxes in Norway and rabbits in Italy (Eisler, 1995). However, it was in upland Britain that deposition of caesium isotopes resulted in contamination of sheep-grazing areas sufficient to cause economic effects on agriculture. The effects were most clearly felt in the uplands of north-west England and in Snowdonia. The UK standard for meat is 1,000 Bq/kg and this was exceeded in sheep that had grazed on contaminated areas. Vegetation concentrations exceeded 6,000 Bq/kg. While transfer of herds to uncontaminated lowland pastures did help them to remove caesium from their tissues through excretion, controls over the sale of meat from affected herds had to be maintained for a number of years following the accident.

Radionuclides in the marine environment

There are a wide range of sources of radioactivity in the marine environment. Any radionuclides that are discharged to the atmosphere can, of course, be deposited into the sea, as with other air pollutants (see Chapter 5). There are also direct sources. Some nuclear waste was routinely dumped at sea. The aim was to dispose of low- and medium-level waste to deep parts of the oceans, where decay rates for the containment materials would be slow and outlast the lifetime of the shorter-lived radionuclides. However, such disposal generated extensive criticism and generally has been suspended. In some areas, however, such disposal was not properly controlled. For example, there is considerable concern currently about the Arctic seas around northern Russia, where obsolete nuclear submarines have been disposed using poor procedures, and elevated radionuclide levels are beginning to be monitored.

In Western Europe the most likely source of routine contamination is, however, from cooling water discharges from nuclear power stations. The most publicised of these has been around the Sellafield nuclear-reprocessing plant in Cumbria, which discharges to the Irish Sea. A range of radionuclides have been released, including relatively long-lived elements such as caesium-134 and 137, ruthenium, strontium-90 and americium. It should be noted that discharge levels were very many times higher in the mid-1970s than today. However, this does mean that the Irish Sea is more contaminated than many other maritime areas and, as a result, can be a source of debate between the Irish and British governments.

Many of the comments made above about the effects of radionuclides in the terrestrial environment are also applicable in the marine environment, e.g. the relative sensitivity of different organism groups. The behaviour of radionuclides in marine systems varies enormously. For example, two decay products of uranium are lead–210 and polonium–210. Lead–210 has no direct biological function and is not a counterpart to a non-radioactive nutrient. It is therefore distributed in the water column (and lost to sediments) predominately by physical processes. Polonium (an alpha-emitter), however, is readily taken up by living organisms. When it enters surface waters, therefore, it is much slower to reach deeper zones, as the element is 'scavenged' biologically. It has been noted (Cherry and Heyraud, 1981), for example, that the shrimp *Gennada valens* in the northern Atlantic can concentrate polonium to the point that it receives a radiation dose equivalent to double to lethal limit for humans.

In assessing pathways to the human food chain, the behaviour of different species is very important. For example, many planktonic species do not appear to accumulate large quantities of radionuclides, as they seem to readily release them back into the environment. However, those species that feed in the bottom sediments (e.g. some crustaceans, flatfish, etc.), may accumulate higher doses. There are, however, important differences between what may be perceived as similar species. For example, lobsters accumulate relatively higher rates of technetium–99 as compared with crabs. Indeed, levels of this isotope in lobsters have risen forty times since 1993 and can now exceed the EU standards for contamination of food following a nuclear accident! Finally, filter-feeders, such as molluscs, are liable to accumulate the highest quantities of radionuclides, just as they do for most marine pollutants (see Chapter 5). It is important, therefore, for the correct components of the marine environment to be monitored in order to assess the likely threat to human health. This requires an extensive, yet targeted, monitoring programme.

The UK has a wide-ranging monitoring programme for radionuclides in the terrestrial and aquatic environment, although here only those within marine areas will be briefly outlined. Table 6.5 outlines the types of monitoring being undertaken. The details of these would reflect the particular needs around specific plants, e.g. shrimps are an important commercial fishery near Hinkley Point power station, so measuring radionuclide contamination of these is necessary. There are also basic environmental

TABLE 6.5   Types of sampling undertaken in routine monitoring in the marine environment for radionuclides in the UK

Measurement	Frequency	Materials or biological material measured
Foods	Monthly–annual	Fish, crustaceans, molluscs, edible plants
Indicator materials	Weekly–annual	Water, sediments, saltmarsh, seaweeds
Gamma dose rates	Monthly–annual	Beaches, harbours, marshes, boats
Beta dose rates	Quarterly–annual	Nets, pots, sediments, saltmarsh
Contamination survey	Monthly–annual	Beaches

Source: MAFF (1996)

or food-chain measurements common to all, e.g. sediment analysis. Some components of the programme are undertaken at relatively frequent intervals, while others, such as an examination of beaches across a wide area, to identify 'hot-spots', is less frequent.

Regulation

In considering the regulation of radioactive substances, it is difficult to separate the principles of the management of individual processes such as power stations from the waste that they produce. Some comments relating, therefore, to waste disposal are appropriate here. However, general waste issues are dealt with in a separate volume in this series.

Most forms of pollution management are types of risk management. Risk management for nuclear-waste disposal and operation of reactors has, however, a very different character to that for other forms of pollution management. This is particularly so if we look to the aims of sustainable development (see Chapter 1), which, in this context, can be interpreted as being that: future generations should be subjected to no greater risk because of the operation of nuclear processes or disposal of the waste than present generations. While for other forms of pollution there may be debate over the risk to future generations in the relatively short term, management of radioactive substances, for example, demands that we analyse the risks for many generations to come, over many thousands of years. The half-life of many of the radionuclides is greater than the period of recorded history, so our understanding of their behaviour into the future is fraught with difficulty.

Let us consider the problems associated, for example, with the creation of long-term underground repositories for high-level waste. There are a range of site-based issues that have to be addressed. The most important of these are the stability of the rocks and their hydrology. Operational issues include matters such as the stability of the containers in which the waste is deposited. Our understanding of these issues is, however, drawn from short-term experience. It is very difficult to be reasonably certain of predicting the hydrology of a site 10,000 years into the future. We have also had insufficient time to perform long-term experiments on the behaviour of the disposal canisters under the elevated temperatures of disposal.

However, a new element for risk assessment in nuclear waste disposal is that of the extreme event, such as seismic activity or meteor impact. These are rare events, but, over the extremely long periods of the life of a disposal site, their importance grows. Thus the probability of a seismic event at a site, for example, of one in a million in a year, becomes one in a hundred over the space of 10,000 years.

It is also important to note that calculations of risk to disposal and operational facilities often do not include the human element. Human error is important. The accident at Chernobyl, for example, was not caused by system failure, but by human error. How can we add the human element to the management of radioactive substances in the long term? It is not possible to imagine what human society will be like in a thousand years' time. It is not, therefore, possible to guarantee that the integrity of disposal sites will not be breached through human activities such as mineral operations, war or terrorism.

In attempting to deal with the long-tern view, the UK National Radiological Protection Board, which advises the government over standards to be applied to protect human populations, issued new advice in 1992 on the safe disposal of waste. This had three main elements:

♦ Future generations should not be subjected to risks that would be considered unacceptable today.
♦ The risk from one disposal facility to the group of people most exposed to it should not exceed a risk of serious health effects to the individuals or their descendants of 1 in 100,000 per year.
♦ The risks to members of the public should be as low as is reasonably achievable (ALARA), taking social and economic factors into account.

It was noted above that it is thought that there is no acceptable safe radiation dose, and that increasing doses of radiation lead to increasing effects. It is not, therefore, possible to establish threshold exposures and standards, as with many other pollutant problems. The management of risk is to determine what is acceptable and what is unacceptable – an assessment based as much on social values as on science. Very clear principles to identify and manage such risk have been formulated by the International Commission on Radiological Protection (ICRP), and have been adopted, for example, by the UK National Radiological Protection Board (NRPB). The ICRP principles involve the following:

1   No process releasing radiation shall operate unless the benefit to the individuals that are exposed or the benefit to society as a whole outweighs the disbenefit to those who are exposed. This is the *justification* for the process.
2   For any source within a process, the size of doses, number of individuals exposed, etc., should be kept as low as reasonably achievable, taking account of economic and social factors. This is the *optimisation* of protection.
3   The exposure that individuals receive from the operation of the process (and neighbouring processes) should be set within specified limits to ensure that no-one is exposed to unacceptable radiation levels. This sets individual risk *limits*.

The key stage in this risk assessment is to determine what levels of exposure are generally acceptable. This is immensely difficult to justify to the potentially affected population. Note that principle 1 above weighs the benefits of society as a whole against small risks to individuals. Principle 3 should protect against excessive doses, but overall it is deemed acceptable if a small risk occurs for a local population as long as there are wider social benefits. While 'on paper' this might seem a reasonable course of action, it is extremely difficult to convince the local population that might be affected. It is not surprising, therefore, that emotions do run high on these issues. Occasionally, the local population might weigh up their own benefits and disbenefits, for example by setting radiation risks against increased employment opportunities.

Natural background radiation provides a dose to the general UK population of 2.2 mSv/yr. It is also important to note that medical exposure results in a further 0.4 mSv/yr. Currently, both the ICRP and NRPB recommend that exposure to the public from man-made nuclear processes should not exceed 1 mSv/yr as a general principle. However, in order to achieve this, the target for the regulators is stricter, to allow for a margin of error. Thus, for a new radiation source, the dose limit should not be greater than 0.3 mSv/yr. This potential increase in exposure above the natural background is less than the degree of variation found in background radiation across the UK, even if radon exposure (see p. 208) is excluded. For existing processes the aim should also be to meet the 0.3 mSv/yr target, although occasionally this may not always be possible.

Whenever there are proposals for a new process, an assessment needs to be undertaken of the risks to the environment. The Environment Agency in England and Wales requires a series of procedures to be followed, in order to ensure that the details of the process operation and the risks associated with it are fully accounted for. An assessment requires an examination of all potential pathways of release, detailing the particular isotopes involved. It is then possible to model how these isotopes would disperse in the environment through air and water as well as through various food chains (e.g. crops, fish, drinking water, etc.) to humans, and to indicate the likely concentrations in these items. It is important then to assess the behavioural aspects of the community, which must be considered to be as normal as possible. For example, if there is marine discharge, do many people swim, eat shellfish, etc.? Obviously, a population is not uniform, and it is possible to divide it into different groups defined by behaviour. Thus children may behave very differently to adults and by their more frequent outdoor activities receive higher exposure. The section of the population that is considered to receive the highest dose is termed the 'critical group'. In the UK, it is generally assumed that if the full assessment is made for local populations, and that process operation procedures will protect people, then the natural environment will be adequately protected. However, as described earlier, this is not necessarily true.

Examples of estimated doses are provided in Table 6.6. This shows the doses estimated from different radiation sources for a collection of English nuclear power stations, assuming that in each case the releases are at the maximum that is permitted. For most instances, there is not a subdivision of the population into different groups. However, for some gaseous discharges this has been done. Thus children are always seen to receive a higher does than adults.

In the UK the keeping, use and disposal of radioactive substances is regulated under the Radioactive Substances Act 1993. All users of radioactive substances require a licence from the Environment Agency (or SEPA, in Scotland), which can issue enforcement, prohibition and revocation notices to ensure compliance with criteria laid down when the licences are issued. If there is an imminent risk of pollution or harm to human health, then the Agency can withdraw the authorisation in whole or part.

A commercial nuclear operation also requires a site licence from HM Nuclear Installations Inspectorate (a part of the Health and Safety Executive), who consult

TABLE 6.6 Calculated critical group doses for the major nuclear sites in England assuming releases are at 100 per cent of those which have been authorised

Power station/site	Critical group doses (mSv/yr)					
	Gaseous discharges			Liquid discharges	Combustible waste	Direct radiation
	Adult	Child	Infant			
Hinkley B		60		5.6	5.2	10
Hinkley (total site)		150		31.9	5.2	100
Dungeness B		6		5.2	n.a.	10
Dungeness (site)		80		14.4	1.63	600
Hartlepool	7.06	8.14	8.04	63.7	0.18	10
Heysham (site)	17.3	33.6	33.6	36.22	0.97	20
Sizewell B	5.7	6.8	15.1	3.3	1.07	3
Sizewell (site)	108.0	119.0	121.0	16.0	2.14	53

Source: HMIP (1996)

the Environment Agency and the Ministry of Agriculture, Fisheries and Food on conditions of operation (Nuclear Installations Act 1965). The Inspectorate may be considered to be most concerned about issues such as workforce exposure to radio-nuclides (protected under the Ionising Radiations Regulations 1985), while wider pollution is more of a concern for the Environment Agency and MAFF.

The peculiar case of radon gas

Most aspects of pollution management, and especially the management of radioactive substances, are aimed at controlling a man-made problem. However, there is one extremely important natural radiological source that poses a widespread threat to human health. Many rocks contain small quantities of natural radioactive elements, such as uranium. There is so much uranium in granite, for example, that it is argued that general exposure to radioactive substances in Aberdeen (the 'granite city') is higher than for those who live around the Sellafield reprocessing plant in Cumbria. However, not all of the radioactivity stays where it is. Uranium decays to radium and this breaks down and produces radon gas. This gas is able to move slowly through rock strata and then out into the atmosphere. Radon is one of the inert gases, so it does not react chemically with the minerals in the rocks and soils through which it passes. For general atmospheric concentrations this poses no threat whatsoever, as the rate of release is very small. However, if the gas seeps into contained areas where it cannot easily escape, then it is possible for concentrations to build up to levels that may pose a threat to health. Such contained environments are commonly found within the foundations and basements of houses and much effort is being put into identifying and managing the resulting risks to health. Of course, not all contained areas need to be artificial. For example, in areas with high radon levels, examination of badgers killed in road accidents has shown that older individuals do show evidence of lung damage consistent with radon exposure. This has presumably occurred due to build-up of radon gas within their underground setts.

The discovery that radon gas poses a threat to domestic residents has a curious history, which is described in greater detail elsewhere (Radon Council, 1995). It involves a member of staff, Stanley Watras, who in 1984 worked at the Limerick nuclear power plant near Philadelphia in the United States. As is common with nuclear plants, each staff member was monitored for radioactive contamination at the beginning and end of every shift. However, Stanley Watras was found to be more radioactive when arriving at work than after leaving. Investigations found that he had bought a house that unfortunately straddled a 10 m wide strip of uranium ore, and that it was highly contaminated with radon gas, previously only thought to be a problem for some miners. The concentrations in this house were very high (around 100,000 Bq/m^3), but wider surveys showed that many homes were at risk.

Radon is a pollution problem that is responsible for about 2,500 lung cancer deaths per year in the UK (about 5 per cent of the total). However, it rarely receives any publicity and, even when information is available, there is great apathy about testing and management. This is extraordinary, considering the public's concern over releases from nuclear power stations and reprocessing plants or concern over the health effects from some other pollution sources described earlier in this volume, which actually pose far less of a risk. Table 6.7 illustrates the relative risk from different radiation sources in the UK for different types of people, including those most exposed to radon (inhabitants of Cornwall) and those working in the nuclear industry, eating seafood in Cumbria (with contamination from Sellafield) or taking frequent air flights (i.e. being exposed to cosmic sources). It can be seen that radon represents a major source in all cases. It is possible that the difference in public reaction is due to the fact that radon is natural. However, it is a salutary lesson to the pollution manager that it is not always easy to engage the interest of the public when a major pollution threat to health occurs.

In 1990 the UK National Radiological Protection Board fixed a standard for exposure for radon of 200 Bq/m^3 as an annual average. This is, of course, not a threshold for effect, as, with any radioactive substance no thresholds can be determined. However, it is a reasonable concentration which, if found in the home, should lead to positive action. For this reason it is not termed a 'standard' but an 'Action Level'. In the US the Environmental Protection Agency has set the value at 150 Bq/m^3. For comparison, a background level of radon is generally about 4 Bq/m^3, while, occasionally, severely affected houses can show concentrations above 2,000 Bq/m^3. Of course the Action Level may be breached in any individual home. However, it is also important to identify those regions of the country where exposure is greatest, in order to focus education and management. Thus surveys are undertaken to determine the degree of radon exposure. Regions of Britain where it is

TABLE 6.7 Differences in the percentage contribution of different radiation sources to the total dose of radiation for different types of people living in the UK

Source	Average person in the UK	Average person in Cornwall	Heavy consumer of Cumbrian seafood in 1987	Average worker in the nuclear industry in 1987	Frequent air traveller
Radon	47	81	39	27	41
Thoron	4	1.3	3.4	2.2	4
Internal	12	4	15	6.6	10
Cosmic	10	3.2	8.7	5.5	22
Fallout	4	0.1	2	0.2	0.4
Occupational	0.2	<0.1	0.2	44	0.2
Discharge	<0.1	<0.1	12	<0.1	<0.1
Medical	12	4	10	6.6	10
Gamma	14	6.3	9.4	7.7	12
Miscellaneous	0.4	0.1	0.3	0.2	0.4
Total dose (mSv)	2.5	7.8	2.9	4.5	2.9

Source: Clarke and Southwood (1989)

considered that present or future houses have a probability of 1 per cent or more of containing radon concentrations above the Action Level are designated as 'Affected Areas'.

Currently the UK has still not completed its survey of radon exposure, with parts of Scotland and Wales yet to be covered. Surveys are simple to undertake. They merely involve the installation of a small passive detector in a house for three months. Care has to be taken to ensure that the data collected are comparable, as factors such as temperature and ventilation have significant effects on radon concentrations. Some of these surveys are undertaken in a systematic sampling pattern, others because of requests from householders. This does mean that survey results are not evenly distributed. While early surveys produced maps based on administrative boundaries (e.g. counties), the threat to houses is based on geology, and so recent surveys based on sufficient data have been able to reinterpret the Affected Areas on this basis (NRPB, 1996). The most seriously affected regions are those granite-based areas of Devon and Cornwall, where up to 30 per cent of houses may be at risk in particular hot-spots. However, it is only possible to provide a statistical probability of the risk of exposure. Individual houses on the same street will vary enormously in the degree to which they are affected, and the only way to assess this properly is to undertake a house-by-house survey. Similarly, granite areas in the Cheviots are at risk, but those in the Lake District do not seem to be. A major band of radon contamination is also found across England, in the Jurassic limestones, ironstones and sandstones that run from Somerset through Gloucestershire and Northamptonshire and into Lincolnshire. Overall it is estimated that 100,000 homes are at risk.

If a house is found to contain high radon levels, then it is important that action is taken. This simply involves the removal of the radon to the outside atmosphere and can be achieved by the installation of ventilation equipment. Obviously the costs of this will vary depending on the layout of a house, but they generally fall within the region of £300–800. There is also a need to ensure that new homes are built with adequate ventilation, to ensure that the problem does not arise. Building regulations have been especially adapted for Affected Areas in order to ensure this. Some would go even further than this. For example, the Chartered Institute of Environmental Health wants the UK government to require that existing home-owners deal with any radon problem in their homes prior to them being sold. This would prevent unwitting purchasers buying a dangerous health hazard.

Chapter summary

♦   Radioactivity results from the decay of unstable elements. Some of these are naturally occurring, and can lead to pollution problems (e.g. radon gas). However, most concern arises from those generated by nuclear power stations or nuclear weapons. Current ambient exposure is low on a global scale. However, hot-spots do exist (e.g. Chernobyl or around old weapons-test sites) where exposure levels are dangerously high. Nuclear reactors and weapons release radiation through their operations and disposal of waste. The latter issue is, however, not dealt with in this volume.

♦   Severe exposure to radiation can cause immediate accute effects on health, e.g. burns or even death. These are predictable deterministic effects. Lower levels of exposure increase the likelihood of cancers. These are stochastic effects, as effects on any individual cannot be predicted with certainty. In assessing health impacts, the routes of exposure and entry to the body are very important, e.g. the lungs are very sensitive. Many radio-isotopes behave like particular non-radioactive elements and this will determine the types of effect that they will have.

♦   In the wider environment, radioactivity affects many plants and animals, though there is a great deal of variation in the degree of sensitivity. It is clear, however, that measures to protect humans alone cannot be considered sufficient to protect the whole environment.

♦   The case of Chernobyl is examined in detail, with long-term impacts on the populations that were exposed to the radiation and to the natural evironment remaining in the excluded area. Effects of the fallout have also had conse-quences as far away as Sweden and North Wales. The management options that were adopted to control exposure are discussed.

♦ Regulation of nuclear power stations in the UK and, generally internationally, is very strict. There are formal assessment procedures determining the likely exposures of different sections of the local population, and releases are regulated in the light of this.

♦ The control of the natural pollutant, radon gas, which derives from the decay of uranium in some rocks and collects in homes, is interesting. Its detection and control are simple and yet there is public apathy towards its management. This contrasts to the very extreme reactions to exposure to much lower levels of some man-made sources.

Recommended reading

Coughtrey, P.J., Bell, J.N.B. and Roberts, T.M. (1983). *Ecological Aspects of Radionuclide Release*. Blackwell, Oxford. Now an older book, this volume still contains a wide range of studies examining how radionuclides are transferred in the terrestrial and aquatic environments.

MAFF (1996). *Radioactivity in Food and the Environment, 1995*. Ministry of Agriculture, Fisheries and Food, London. This report (and subsequent ones in future years) presents the detailed results of the UK radiological monitoring programmes. These monitor general radioactivity (e.g. from Chernobyl) and also specific programmes around individual power stations and other installations. It provides detailed data on exposure levels, and is an excellent source book.

Radon Council (1995). *The Radon Manual*. The Radon Council, Shepperton. This is a complete guide to the problem of radon pollution. It explains the origin, detection and management of radon in detail, although much is written for a non-technical audience.

Savchenko, V.K. (1995). *The Ecological Effects of the Chernobyl Catastrophe*. Man and the Biosphere Series No. 16. UNESCO, Paris. This is the most comprehensive account of the Chernobyl accident and the local and international effects that took place subsequently.

Chapter 7

People, politics
and the press

◆ Introduction   214
◆ The public   215
◆ The media   219
◆ Politicians   221
◆ Industry   222
◆ Non-governmental bodies   223
◆ The anti-environmental backlash   224
◆ Chapter summary   225
◆ Recommended reading   225

Introduction

The work of pollution management cannot be undertaken in isolation from the sociopolitical context in which it is set. This chapter will provide a brief overview of this. It will examine the role of the public, media, politicians, industry and non-governmental organisations in managing pollution and reacting to those who undertake such management. In particular, the chapter focuses on the issues that tend to motivate each of these groups and the ways in which communication between them and the pollution manager can be made easier.

There are many ways in which the environmental manager is dependent upon and is able to influence the society in which he or she works. These include:

♦   The regulatory framework, which has been created over many years to reflect the changing concerns of society as much as scientific developments, and which is the basis of action for most pollution management decisions.

♦   The regulatory framework is created by politicians. Pressure relating to individual decisions may sometimes reflect the different interests of politicians, e.g. the short-term vision and either national or constituency concerns of politicians may not always lead them to support every individual decision.

♦   Pollution managers may act to settle disputes between opposing land management options (e.g. the interests of a farmer and those of the local angling club). While there are options for satisfactory conflict resolution in some instances, this is not always the case, and an aggrieved party may create adverse publicity, etc., for the pollution manager, or even resort to legal action.

♦   Pollution managers may at times need the support of the media to promote the results of research, policies or individual decisions to gain wider support for their work. However, the media will want a story that fits their needs and so may offer a different perspective to that desirable to the pollution manager.

♦   Adverse media coverage is also a big problem. This may be from sections of the media that are hostile to environmental issues and present attempts to control pollution as 'bureaucrats stifling the efforts of industry', or from the opposite end where possible sensible compromises are portrayed as 'a betrayal of the environment' and that a regulator is 'lacking any teeth.' It is a common problem among all regulatory agencies that they walk the tightrope between being billed as a 'watchdog' or a 'lapdog' and even, occasionally, both at the same time!

These days much fun is made of 'management speak' within public agencies and government departments. However, it is important to note that the public, journalists and politicians who – in theory – serve it, are the 'customers' of the pollution manager. Of course, one may argue that the wider environment, e.g. wildlife, is the 'customer', and this is also true. A consideration of the relationship between the pollution manager and public as 'customer–provider' is, however, a useful one, as it not only enables a clearer focus on work priorities, but also addresses many communication issues. The difficulty with this model of the relationship is that the customer is not 'always right', as can be argued in commercial relationships. The public, press or politicians may not understand the issue at hand, or have very different objectives. While it is imperative that every effort is made to increase that understanding, the professional status of an environmental manager may mean that they have to take very unpopular decisions. Hopefully, a better emphasis on communication should reduce such situations to a minimum.

The public

The role of the general public should never be underestimated. It is true that the 'silent majority' are often silent. However, if sections of the public are sufficiently motivated, then they can have considerable influence. This can, of course, work in both positive and negative ways for pollution managers. For example, a pollution manager may act to close down an industrial process that has led to air quality standards being exceeded for nitrogen dioxide in a local community, and that is unable to clean up its act. How will the public react? There has been much publicity about air pollution and its role in causing asthma, and it is possible that the local community may be very supportive of the decision. However, if that community is dependent on that process for jobs, then the reaction may be quite different.

It is important to note that literally millions of people are members of environmental organisations. Angling and walking are by far the most popular outdoor pastimes and provide many people with at least a casual interest in the environment. Of course there is overlap between membership, and some members may not be particularly committed to 'the cause'. However, it illustrates a deep general concern for the environment that a pollution manager should tap into when it is useful, and be aware of if his or her decision is liable to be unpopular.

While a pollution manager has to make a decision based on sound science and within the correct statutory framework, and therefore may have little room for manoeuvre, that person must take account of local public interest in the way in which the decision is announced.

A key area of concern in dealing with the general public is the presentation of technical information. Pollution management is highly technical. It is difficult to discuss air pollution impacts from an industrial process on human health, for example, without an understanding of engineering, atmospheric chemistry and physics, and epidemiology. Few pollution managers are experts in all these fields and yet any decision has to communicate some part of these to the general public. What is harder still is that very often such communication is not direct, but is via the media. However, the media are a separate issue and will be considered on p. 219.

In looking at technical information and its presentation, it is important to focus on the key issues within it. There is little point in attempting to transmit a host of numbers or conclusions. A good way of focusing on the key point of the issue is to write its conclusion in one sentence (without too many subclauses). Having done that, individual components can be expanded upon to provide more information for those that require it. This is a way of writing or speaking very different to that which most pollution managers are used to. Pollution managers are often scientists or engineers by background. In presenting information in these fields, one generally begins by setting out a problem, describing the work undertaken to assess it and ending with the conclusion. In dealing with the public (and the same can be said for media and politicians), this process should generally be reversed – begin with the conclusion.

It is important to be very careful about how information is worded. Take, for example, a hypothetical case of a change in use of fuel at a cement kiln to one derived from chemical waste. Such fuel will emit small quantities of dioxins. Dioxins are not only highly toxic, they are recognised as such by the general public. A major problem for the pollution manager is that, if the use of such fuel is deemed safe, the dioxin emissions have to described as safe. This is a terribly difficult thing to do, as technically there is no strictly safe level. However, presentation can help. One could state, for example, 'studies have shown that there is no evidence that the dioxin emissions will cause any harm to the health of the local population.' This may be a perfectly correct statement to make, but it is immediately open to the charge that either the pollution manager has not looked hard enough for evidence or, as many journalists and some members of the public will respond, by saying, 'but you have no evidence that it is safe.' While the pollution manager does in fact have such evidence, time and again the response is hedged with so many caveats that the decision becomes open to ridicule. It is better to state that 'dioxin emissions have been shown to be below levels which can cause any illness in people.' This puts the presentation on the offensive and the most likely response is to seek clarification.

It is important to remember that those most affected by a decision will be the most vociferous. They may not, however, be the most important. In the UK at present this is typified by responses to a number of road schemes, where pollution issues play a role, although it is usually not a central one. The general scenario is of local concerns over traffic or the environment against national concerns over traffic or the environment. The opposition can be in either direction. However, there may be important, national environmental reasons for not agreeing to local pressure and this is a difficult situation for an environmental manager to cope with.

It is important to reach out to the public at frequent intervals. This may be easier to manage if a pollution manager is operating in a distinct geographical area, but consideration to this should be given by all. It is better to discuss some issues with the public when attention is not focused on a particular decision. It may be useful to meet public-interest groups to explain the work of the organisation for which one works, or to outline a particular pollution issue. For example, one may discuss pollution impacts with angling groups or health impacts with local medical professionals. This exchange of information and views then improves communication when conflicts do arise, and may help resolve them more easily.

PLATE 7.1 Protests against polluting activities take place all around the world. This photograph from northern Spain illustrates concern about nuclear power.

Photograph: Andrew Farmer.

Involvement of the public is also a key element in achieving sustainable development. An environmentally sustainable economy will require changes in lifestyle and support from the community. It is important, therefore, for pollution managers to communicate issues in this wider context. It is also important that people recognise what is of value in their own environment, and how they want that environment to be protected and enhanced. This may be best achieved through various processes under the umbrella of 'Local Agenda 21'. The issue of pollution and what can be done about it, is a central topic in many LA21 discussions.

CASE STUDY: SCIENCE, LAWYERS AND MONEY FOLLOWING THE EXXON VALDEZ OIL SPILL

In 1989, the Exxon Valdez oil tanker spilled 37,000 tonnes of oil into the Prince William Sound in Alaska. This is an area noted for its extremely rich and important wildlife. The immediate reaction was one of horror, and soon thousands of birds and mammals were found to have been killed. There was even greater concern over the long-term impacts of the spill and pressure to get the oil company, Exxon, to pay compensation for the damage.

Exxon lost heavily in monetary terms. First, they were required to pay for all of the clean-up operations, amounting to US$2.5 billion. Second, they were required to pay $900 million to a panel of state and federal trustees to pay for future restoration, and pay $39 million to fishermen. The trustees will also receive a further $100 million if, by 2006, further – as yet unidentified – oil damage is discovered. Finally, and most significantly, a court case imposed a punitive fine on the company of $5 billion.

Given the enormous sums of money involved, it is not surprising that the relationship between science, the public and the legal system became confused. A range of organisations, from environmental groups and the state, through to Exxon itself, gathered scientific information for legal and other proceedings. Scientist was set against scientist. This did not aid communication with the public. For example, at the end of the trial for punitive action, one juror remarked:

> 'You got a guy with four Ph.D.'s saying no fish were hurt, then you got a guy with four Ph.D.'s saying, yeah, a lot of fish were hurt . . . They just kind of delete each other out.'

This is a common feature of many exchanges at legal proceedings or during public inquiries where scientists and other experts will be ranged against each other. Obviously, in this case, the stakes were extremely high. Court proceedings may also be very different if the decision is to be made by a jury. In this case, the ability of a lay juror to cope with specialist information may be much less than an experienced judge or inspector.

The settlements for Exxon Valdez have also affected the way that science has been conducted subsequent to the trials, etc. In particular, the company has the potential for a further $100 million payout, if damage, currently unknown, is discovered. Thus the way in which scientific research and monitoring is undertaken is of continuing financial interest. Much of the work done by different groups of scientists has been kept secret, with little dialogue or data sharing. Thus millions of dollars have been spent duplicating work and much of the programme has come in for scathing criticism.

The biggest lesson of the Exxon Valdez has been that a major company can expect huge financial implications for causing environmental pollution. This has sent a number of shock waves through companies examining their risk management procedures and insurance policies. However, it has also been a salutary lesson in examining the way in which scientists conduct themselves and the way in which any such clean-up operation could be managed in future.

The media

Television, newspapers and radio tend, if this is not a contradiction in terms, to receive quite a bad press. This may be so particularly among professional people, who may feel they have been treated badly by the media. However, the importance of the media should not be underestimated. Opportunities for using the media should be taken and potential threats recognised.

Most journalists are just journalists, with no particular axe to grind with regard to environmental issues. There are a few, however, with particular agendas. These occur on both pro- and anti-environmentalist sides. An anti-environmentalist journalist may be unlikely to portray a pollution management decision in a favourable light. However, while a pro-environmentalist journalist may be looked at as a 'friend' rather than a 'foe', their agenda may be such as to distort what may being presented. For example, the journalist may overemphasise the conclusions that have been reached, and may not allow for 'qualifications' to any statement.

Journalists are usually not interested in a story if they think that the public will not be interested. A programme of work to upgrade sewage works to improve water quality is likely to get coverage if the benefits to otters are highlighted, but unlikely to be covered if the benefits to mayflies are stressed. The role of the 'cute and cuddly' in wildlife stories is well established, and presentations to journalists should account for this. Similarly, health effects on children tend to be more newsworthy than similar effects on adults.

Journalists want a story that is focused on their own market. The presentation of a pollution manager may not be synonymous with this, and, as a story unfolds, distortions may result. It is also worth noting that television or newspaper photographers may require quite specific images that may not be easy to obtain. For example, after the 1995 Sea Empress oil spill in Wales, stories emerged of reporters arranging for healthy sea-birds to be dipped in oil in order to get the picture that needed to go with the story. Such a picture might actually aid a pollution manager in achieving what he or she wants, but it is hardly ethically justified.

There are many more extensive fora for providing advice on how to deal with journalists. However, there are a few points worth making.

- ◆ Providing information on a press release is useful. However, it is important to realise that journalists will receive many press releases every day. Therefore, if a pollution manager wishes a story to be covered, the first words (title or first sentence) must be eye-catching. A press release that begins 'Agreements have been reached to upgrade the sewage works at . . . ' will be most likely to be in the bin before the sentence is finished, but one that starts '£2 million to be spent to save otters . . . ' may receive some attention. The otters may not be central to the story, but they may be the vehicle for its publicity.

- ◆ Be helpful and be ready to provide additional information as needed. For example, have maps and figures or black-and-white or colour pictures available. Improving the presentation of a story will encourage journalists to use it.

- ◆ Few stories are appropriate for all media. If the story is a local one, stick to local media, otherwise other journalists will become used to receiving 'irrelevant' press releases. It is also worth scanning newspapers in particular, for the types of story that they cover, and focus attention on these in future. If those papers have 'environment' sections on particular days, or the television or radio have particular programmes to be targeted, then you should time the release appropriately.

- ◆ Be careful to avoid days likely to be filled with other news (e.g. the budget or even a local carnival). If possible, target slack periods, e.g. around holidays, when journalists may be more desperate for news.

- ◆ In presenting information on radio or television, particularly in live interviews, it is important to organise facts and opinions into small sentences. Most interviews are very short and my be further edited for repeat broadcasts. Politicians may be criticised for the use of the 'soundbite', but this is only a response to a demand from the media. If facts are put in a 'nutshell' of a few seconds, then they are more likely to be used.

- ◆ It may also be worth 'cultivating' one or more appropriate journalists if regular contact is likely. This may include the option of an occasional 'exclusive'. Not only may this allow a clearer presentation of the pollution manager's own story, comments may also be sought by the journalist on other stories that may be relevant and for which a public view is useful.

- ◆ Unwelcome press attention is difficult to deal with. A 'no comment' response is useful, as there is not a lot that can be done with it. It is, however, better to organise the facts and present an alternative view if possible. Again, it is important to be positive, i.e. to present it as one would from scratch, not as a defensive response to criticism from another.

These are only a few pointers to relationships with the media. However, they give a flavour of the delicate balance between good publicity, bad publicity and none at all, and between good and bad relationships with journalists. A pollution manager is likely to have to manage the press at some point in his or her career. Even in organisations with professional press officers, the pollution manager is likely to have to present the story, albeit with help, as only he or she may have the technical background to answer questions. It is extremely unlikely that a pollution manager will spend his or her entire career making decisions or producing research that no-one is interested in.

Politicians

Many people involved in pollution management are directly or indirectly responsible to politicians, either national or local. All regulatory agencies operate on behalf of governments, and have to act primarily within the statutes laid down by those governments, and secondarily within other policy frameworks. For those not responsible to politicians, their most likely working environment is within a non-governmental body that has a key aim of influencing government policy. For both groups, the relationship with politicians is crucial, though of a different kind. Those within the machinery of government will need to use official channels of one sort or another, while a wider group of options is available to those outside of government agencies.

To paint all politicians as self-seeking publicists who do not look further than their next election is to paint a caricature which, while it may be true for some, does do an injustice to those who have genuine concerns and commitment. The difficulty for pollution managers in any situation is that these two extremes of politician could turn up at any point within governments, and the approach needed for each may be quite different.

The politician most concerned with short-term interests is best influenced by reference to the views of the public. If concern among the public for their health or concern for the wider environment is increased, then sometimes a great deal can be achieved by stressing the importance of an issue to a politician's constituents. Green issues wax and wane in importance for the general public, and, as a result, the time that politicians will give to them also varies greatly. However, while green issues in general may not, at one time, be particularly prominent in the news, this does not prevent a specific issue receiving sufficient publicity to grab political attention.

Communication with politicians requires considerable skill. They need more information than would be presented, for example, to a journalist, but they do not have the time to digest large technical reports. Well-written summary documents are exceedingly useful, in that they enable a rapid absorption of the main points of interest and can provide further background detail if the reader is interested. Indeed, consultancies exist to seek out such technical information and summarise it in forms that politicians can use.

In communicating with a policy-maker within a government department, it is generally straightforward as to who to contact. However, there are times when a more widespread coverage of members of legislative bodies is appropriate. In this instance, sometimes it is not a wise idea to provide information to every member, but to target either those with a known interest in environmental issues or those with other concerns relating to the subject at hand.

In order to influence government policy regarding pollution management, it is important to be quite precise as to who has control over the issue that is being addressed. For example, there is little point (except to gain additional support) in lobbying a local authority in Britain over emissions from a power station. They do not have authority over it. Of particular difficulty is the role of the European Union in pollution regulation. While the remit of current Directives is generally clear, future legislation is not limited to these areas alone. If one country alone is resisting pollution control legislation, lobbyists within that country can sometimes achieve national objectives by seeking change at a wider European level.

It is also important to note that policy-making is not undertaken by politicians alone. In all countries they are supported by an extensive civil service. The exact relationship between the civil service, its Ministers and agencies varies from country to country and even, in subtle ways from department to department within one country. Agency pollution managers need to work through the departmental civil service, who should themselves take pains to seek opinions from those agencies on all relevant developments and not wait for comments to surface. Non-governmental pollution managers may also achieve much by approaching civil servants. Civil servants are very likely to be more technically versed in pollution issues and will be prepared to work though more detailed presentations. It may also be possible to get a clearer picture of why particular policies are or are not being adopted, e.g. an explanation of technical problems or where difficulties may exist between departments. Any relationship cultivated with a civil servant must be carefully maintained. It is they who advise the Minister on a regular basis, and any breakdown in the relationship (e.g. a breach of confidence) may cause lasting damage.

Industry

It is a serious error to view industry as antithetical to environmental concerns. On many occasions they may be demonstrably more forward-thinking than governments who take restricted action in order to protect industry's 'interests'. Of course there are a number of 'cowboy' operators with no environmental concerns, but these few should not be allowed to characterise the rest of industry.

Robinson (1996) argued that commercial activity and environmental protection are, in the long term, interlinked. In particular, companies should be aware of three variables that affect their survival. Threats may occur in one of these areas or may reflect an interaction between them. Each company should produce a plan for a sustainable business addressing each of these issues, and a pollution manager will have a significant contribution to make to these. These variables are:

1   *Resources.* These include not only raw materials, but the provision of environmental services to act as pollutant sinks.
2   *Innovation.* Commerical activity can be severely threatened by technical advances by other companies that are more environmentally 'friendly' and that may become obligatory.
3   *Values.* The values held by society, especially the companies' customers, are very important, but are often not addressed by companies.

Industry has focused largely on the regulatory and policy framework provided by government, and has taken a view that as long as its actions are consistent with this, then there should be no problems for its operations. However, it is clear that there is a wider context in which companies must operate. This was exemplified by the Brent Spar incident.

In deciding to dispose of the Brent Spar oil platform, Shell communicated with the UK government's Department of Trade and Industry, who examined the scientific basis for the potential environmental effects of the proposed disposal. It was considered that the potential damage to marine life was less than might take place through onshore disposal, i.e. marine dumping was determined as BPEO. While there is now some doubt as to whether this would in fact, have been the case, the real problem over the Brent Spar was that the debate was held in contrasting contexts. Shell based their arguments on the best option for the disposal of an oil platform. However, the public and NGO debate was largely about the whole policy of ocean-based waste disposal. The latter has an additional political context that is not amenable to scientific analysis. The key lesson to be learnt is that companies which deal only with government and regulators and do not take account of the wider concerns of society, may be taking serious commercial risks.

It is increasingly evident that the old areas of confrontation between industry and environmental groups need to be reassessed. There are now good examples of companies and NGOs working together to seek solutions to environmental problems and that are, at the same time, profitable for the company. The search for 'win/win' solutions is essential, although it should not be expected that they will be found in all cases. However, a dialogue between industry and environmental organisations should now be seen as good management practice.

A good example of collaboration between industry and NGOs is that of the Calor Group and Greenpeace. Greenpeace had established an extensive and effective campaign against the use of CFCs in, for example, refrigerants. However, the initial response of industry was to replace CFCs with HCFCs, which have a lower, but still important, ozone-depleting potential. Greenpeace, therefore, persuaded Calor to develop 'Greenfreeze', a hydrocarbon refrigerant that is completely ozone 'friendly'. The refrigerant market was a new venture for the company, so this was a major commerical step to take. The partnership with Greenpeace was maintained, with the NGO using its world-wide network of contacts to promote Calor's product. This included the fastest-growing refrigerant market in the world, i.e. China.

Non-governmental bodies

The term 'non-governmental organisation' (NGOs) covers a wide range of bodies. Some may be geographically local, e.g. a county wildlife trust in England. Some are focused on particular interests, e.g. one dealing with asthma. Others may encompass extensive environmental concerns and be national or international, e.g. Friends of the Earth or Greenpeace. Some of them also act as professional organisations for academic or other workers, e.g. the National Society for Clean Air and Environmental

Protection is one of the oldest environmental NGOs in the UK, but it also acts as a professional forum for local-authority environmental health officers. Overall, the membership of environmental bodies is extensive and greater than the membership of all the political parties combined. In the UK, for example, the World Wide Fund for Nature has about 200,000 members and the Royal Society for the Protection of Birds has 850,000 members.

Few of these bodies (excepting the NSCA) contain personnel actually managing pollution. However, most have a deep interest in how pollution is managed, and may have staff devoted full-time to working on these issues. As with all organisations, each has its own particular interests and philosophies and these must be borne in mind when interacting with them.

The greatest dilemma that tends to face environmental NGOs is over whether compromise is allowable. The conflict between those who stick rigidly to a 'pure' environmentalist agenda and those who are willing to agree to improvements and interim solutions is best seen in the fate of some of the European Green Parties (although these are, of course, political parties and not NGOs). The split between 'fundamentalists' or 'deep greens' and others, has certainly damaged the Green Parties in a number of countries.

NGOs may also set a particular issue or campaign in a wider context, although that context may be hidden from view. For example, if the NGO considers that it is necessary to attack a government on its environmental record, and is using a particular issue to help achieve that, it is unlikely to welcome solutions offered to that problem. They are likely to be portrayed as 'insufficient'. If government agrees to the NGO's campaign, then it is common for little credit to be given to the government, and the NGO's role will be highlighted. This is a natural campaigning technique and publicises the effectiveness of the organisation and helps increase membership. However, if the aim of campaigning is limited to a particular issue, then the potential of working in partnership and seeking common solutions with an NGO is possible. When all NGO activity is considered as a whole, the latter situation is the most likely. However, generally the most publicised cases fall into the former category.

The anti-environmental backlash

In assessing the role of pollution managers and their political context, it is important to note that their work will not always be considered in a rational context. Around the world (though especially in the United States), there is an increasing coalition of anti-environmental interests that will use any means to achieve their aims. In the US, these groups have combined under the umbrella of the Wise Use Movement, whose aim is 'to destroy, to eradicate, the environmental movement' (Rowell, 1996). This is achieved by portraying environmentalists as irrational, as Communists and fascists, by attacking their science with bogus arguments from a handful of maverick academics and, most disturbingly, by using violence. Much of the focus has so far been on environmental issues facing land use (e.g. logging or mining). However, in 1991 the director of Greenpeace's USA science unit was subject to an arson attack because of her work on toxic waste incineration and the pollution that it would cause. In the UK there is, fortunately, not a similar organisation. However, it is possible to see the caricaturing and irrationality, common in the US, from some politicians and newspaper columnists.

This is not the place to consider this issue further, and the reader is directed to Rowell (1996) for a fuller discussion. However, it does have lessons for those working in pollution management. First, it is important to ensure that the scientific basis for action is well founded. If decisions can be undermined on one occasion, then the anti-environmental lobby will use these on other occasions when decisions are more sound. Second, great care needs to be taken over communication of ideas of risk and uncertainty. This has been discussed earlier. However, in this context it is easy to see how anti-environmentalists can use a statement of uncertainty to undermine a reasonable basis for action. Finally, for a few of the people working to reduce environmental pollution, there may be times when courage is required. Let us hope that, in the UK at least, this will be rarely needed.

Chapter summary

The work of a pollution manager cannot be undertaken in isolation from the social and political environment in which he or she operates. This chapter provides a brief overview of some of the factors that may need to be considered, including:

♦   The role of the public: communicating technical information, local or national concerns.

♦   The role of the media: publicising 'good' stories and managing 'bad' stories, timing of communication and presentation of technical information.

♦   The role of politicians: briefing skills, appealing to self-interest, targeting those with influence.

♦   The role of industry: instilling an environmental culture within management structures, encouraging full environmental management.

♦   The role of NGOs: a variety of bodies with contrasting objectives, which are not always supportive of the pollution manager.

Finally, some consideration is given to irrational responses to environmental management, where in a few extreme cases, 'environmentalists' have been subject to vilification or even violent attacks.

Recommended reading

Boehmer-Christiansen, S. and Skea, J. (1991). *Acid Politics*. Belhaven Press, London.
    This is a detailed comparison of how the UK and Germany have dealt with the
    problems of acid rain. It examines the different pollution climate, and sets the

government responses in the light of their respective regulatory philosophies and political and social culture. It also examines the role of science and economics in the respective governments' thinking.

Rose, C. (1990). *The dirty Man of Europe*. Simon and Schuster, London. This is an excellent introduction to a view from a seasoned campaigner within environmental NGOs. The book examines the UK's record in a wide range of areas, e.g. acid rain, nitrates in drinking water, lead, traffic, radioactive waste, litter, pesticides and others. In each case, the author describes in detail his view of why the government has not taken strong enough action. This includes interesting asides on some unlikely influences on policy, and the role that individual personalities may have in the process.

Rowell, A. (1996). *Green Backlash*. Routledge, London. This is the book to frighten environmental managers. It describes the backlash to environmental activists in the US and other countries. Sometimes this can even be violent. The reaction may be far from logical and open to rational discussion.

Glossary

Acid rain: This has a variety of meanings, ranging from rain that has been acidified by pollutants to a general term incorporating all acid deposition and acid gases in the atmosphere.

Acute effects: Symptoms or injury that occurs rapidly, e.g. after exposure to pollutants.

Alveoli: The small air cavities within the lungs.

Benthic: A term referring to the occurrence of life or materials on or near the floor of the sea.

Bioaccumulate: The mechanisms whereby organisms are able to concentrate pollutants that are present in low concentrations in air, soil or water to higher concentrations in their tissues.

Biodiversity: A generic term to describe all aspects of the diversity of life, incorporating issues such as species richness and ecosystem complexity.

Biofuel: A fuel derived directly from a biological source, e.g. ethanol produced from sugar cane.

Bryophyte: A division of the plant kingdom that includes mosses and liverworts.

Carcinogen: A substance that is capable of initiating cancers.

Cation exchange capacity: The total amount of cations that a material (e.g. soil or cell wall) can absorb at a given pH.

Chronic effects: Disease symptoms that develop over a long period of time.

Contingent valuation: A method that asks individuals to set prices on goods, e.g. environmental assets.

Critical level: The concentration of pollutants in the atmosphere above which direct adverse effects on receptors, such as plants, ecosystems or materials, may occur, according to present knowledge.

Critical load: The highest load of a pollutant that will not cause chemical changes leading to long-term harmful effects on the most sensitive ecological systems, according to present knowledge.

Diffuse pollutants: Pollutants that do not arise from specific points, e.g. nitrates leaching from a field.

'Effects-based approach': An approach to pollution control policies that is linked directly with the environmental consequences of such controls, rather than with arbitrary reduction targets.

Encephalopathy: A general disease of the brain, associated with toxic poisoning.

Epidemiology: The study of diseases within a population, identifying their occurrence and the role of environmental factors.

Epiphyte: An organism that grows on a plant, e.g. a lichen on a tree or algae on an aquatic plant.

Eutrophic: The state of a habitat (aquatic or terrestrial) that has excess nutrients. This may be natural or the result of nutrient pollution (eutrophication).

Flue gas: The gaseous products of combustion within a chimney.

Fugitive emissions: Pollutants from a process not arising from specialised release points (i.e. a chimney), e.g. an organic solvent escaping from a sealed area.

Free radical: Groups of atoms in free existence for only short periods as they have unpaired electrons. They are therefore usually very reactive.

Ground-level concentration: The air pollutant concentration at ground level, resulting from chimney emissions – used in monitoring or modelling assessments.

Heavy metals: A loose term to describe metals of high atomic number, usually toxic.

Hydrological source area: A part of a catchment where streams may begin to flow, e.g. spring sites.

Hydrosphere: The total water environment of the Earth, i.e. seas, freshwaters and ice.

Hypoxia: Lack of oxygen supply or stress induced by very low oxygen conditions.

Ion exchange site: A molecular location on matter (e.g. a cell wall), which will exchange one ion for another of the same charge.

Macrophyte: Literally 'large plant', used to describe higher aquatic plants to distinguish them from phytoplankton (microphytes).

Mycorrhiza: A symbiotic relationship between a fungus and a root, from which both fungus and plant appear to derive benefit.

NOEL (No effect level): A concentration or dose below which it is thought that no effects occur on a particular organism or population of organisms.

Non-governmental organisation (NGO): Any organisation not linked to government (usually involved in some form of campaigning), e.g. Friends of the Earth.

Occult deposition: Deposition of mist or cloud water which may be contaminated with pollutants.

Oligotrophic: Of very low nutrient status.

Pathogens: Any organism which produces a disease.

Plume: The pathway taken by pollutants discharged by point sources in air or water.

Prescribed substance: A pollutant that is listed in legislation as requiring controls over its emissions.

Primary pollutant: A pollutant that is released directly to the environment.

Secchi depth: The transparency of water measured by a assessing the visibility of a simple disc lowered into it.

Secondary pollutant: A pollutant that arises due to reactions that take place in the environment, e.g. involving primary pollutants.

Seeder–feeder effect: The effect where rain passes through polluted cloud water and absorbs pollutants.

Synergistic effects: Where pollutants may act together to form effects greater than predicted by summing the individual impacts.

Willingness-to-pay: The measure of an individual's preference for an environmental good, as judged by asking how much they are willing to pay for it.

Wind rose: A diagram showing frequency and strength of winds at a location from different directions over a set time period.

Zero-emission vehicle: A vehicle that does not directly cause emissions (usually an electric vehicle), though technically such energy use may result in emission elsewhere.

References

Abel, P.D. (1989). *Water Pollution Biology*. Ellis Horwood, Chichester.

Anon. (1995). Bones reveal medieval air pollution. *British Archaeology*, 2: 5.

Anon. (1996). Controlling sediment in streams and watercourses. *Enact*, 4 (3): 8–9.

Arnolds, E. (1991). Decline of ectomycorrhizal fungi in Europe. *Agriculture, Ecosystems and Environment*, 35: 209–44.

Ashenden, T.W. and Edge, C.P. (1995). Increasing concentrations of nitrogen dioxide pollution in rural Wales. *Environmental Pollution*, 87: 11–16.

Ashenden, T.W., Bell, S.A. and Rafarel, C.R. (1990). Effects of nitrogen dioxide pollution on the growth of three fern species. *Environmental Pollution*, 66: 301–8.

Ashmore, M.R. and Wilson, R.B. (1994). *Critical Levels of Air Pollutants for Europe*. Department of the Environment, London.

Ayazloo, M. and Bell, J.N.B. (1981). Studies on the tolerance of SO₂ of grass populations in polluted areas. I. Identification of tolerant populations. *New Phytologist*, 88: 203–22.

Baddeley, J.A., Thompson, D.B.A. and Lee, J.A. (1994). Regional and historical variation in the nitrogen content of *Racomitrium lanuginosum* in Britain in relation to atmospheric nitrogen deposition. *Environmental Pollution*, 84: 189–96.

Baker, R.J, Van Den Bussche, R.A., Wright, A.J., Wiggins, L.E., Hamilton, M.J., Reat, E.P., Smith, M.H., Lomakin, M.D. and Chesser, R.K. (1996). High levels of genetic change in rodents of Chernobyl. *Nature*, 380: 707–8.

Bates, J.W., Bell, J.N.B. and Farmer, A.M. (1990). Epiphyte recolonisation of oaks along a gradient of air pollution in south-east England. *Environmental Pollution*, 68: 81–99.

Bates, T.S., Lamb, B.K., Guenther, A., Dignon, J., and Stoiber, R.E. (1992). Sulfur emissions to the atmosphere from natural sources. *Journal of Atmospheric Chemistry*, 14: 315–37.

Battarbee, R.W. (1984). Diatom analysis and the acidification of lakes. *Philosophical Transactions of the Royal Society of London*, 305B: 451–77.

Beebee, T.J.C., Flower, R.J., Stevenson, A.C., Patrick, S.T., Appleby, P.G., Fletcher, C., Marsh, C., Natinski, J., Rippey, B. and Battarbee, R.W. (1990). Decline of the natterjack toad (*Bufo calamita*) in Britain: palaeoecological, documentary and experimental evidence for breeding site acidification. *Biological Conservation*, 53: 1–20.

Bell, J.N.B. (1994). A reassessment of critical levels for SO_2. In M.R. Ashmore and R.B. Wilson (eds) *Critical Levels of Air Pollutants for Europe*, pp. 6–19. Department of the Environment, London.

Bennett, J.H., Lee, E.H. and Heggestad, H.E. (1990). Inhibition of photosynthesis and leaf conductance interactions induced by SO_2, NO_2 and $SO_2 + NO_2$. *Atmospheric Environment*, 24a: 557–62.

Betts, W.E., Floys, S. and Kvinge, F. (1992). The influence of diesel fuel properties on particulate emissions in European cars. *Society for Automotive Engineers Technical Paper* 922190. The Engineering Society for Advancing Mobility in Land, Sea, Air and Space, Warrendale, Pennsylvania.

Bobbink, R. (1991). Effects of nutrient enrichment in Dutch chalk grassland. *Journal of Applied Ecology*, 28: 28–41.

Boehmer-Christiansen, S. and Skea, J. (1991) *Acid Politics*. Belhaven Press, London.

Bower, J.S. (1996). The UK National Urban Air Monitoring Networks. In C.J. Curtis, J.M. Reed, R.W. Battarbee and R.M. Harrison (eds) pp. 21–1. *Urban Air Pollution and Public Health*. ENSIS Publishing, London.

Brown, V.C., McNeill, S. and Ashmore, M.R. (1992). The effects of ozone fumigation on the performance of the black bean aphid, *Aphis fabae* Scop., feeding on broad beans, *Vicia faba* L. *Agriculture, Ecosystems and Environment*, 38: 71–8.

Bull, K.R. (1991). The critical loads/levels approach to gaseous emission control. *Environmental Pollution*, 69: 105–23.

—— (1992). An introduction to critical loads. *Environmental Pollution*, 77: 173–76.

Butlin, R.N., Yates, T.J.S., Murray, M. and Ashall, G. (1995). The United Kingdom National Materials Exposure Programme. *Water, Air and Soil Pollution*, 85: 2,655–60.

Carvalho, L. and Moss, B. (1995). The current status of a sample of English Sites of Special Scientific Interest subject to eutrophication. *Aquatic Conservation: Marine and Freshwater Ecosystems*, 5: 191–204.

CEC (1992). *Forest Condition in Europe. 1992 Executive Report*. Commission of the European Communities, Brussels.

Cherry, R.D. and Heyraud, M. (1981). Polonium-210 content of marine shrimp: variation with biological and environmental factors. *Marine Biology*, 65: 165–75.

Cheung, K.C., Chu, L.M. and Wong, M.H. (1993). Toxic effect of landfill leachate on microalgae. *Water, Air and Soil Pollution*, 69: 337–49.

CLAG (1994a). *Impacts of Nitrogen Deposition in Terrestrial Ecosystems. Report by the UK Critical Loads Advisory Group*. Department of the Environment, London.

CLAG (1994b). *Critical Loads of Acidity in the United Kingdom. UK Critical Loads Advisory Group Summary Report*. Department of the Environment, London.

Clark, R.B. (1992). *Marine Pollution*. Clarendon Press, Oxford.

Clarke, R.H. and Southwood, T.R.E. (1989). Risks from ionising radiation. *Nature*, 338: 197–8.

Dickson, W. (1975). The acidification of Swedish lakes. *Report of the Institute of Freshwater Research, Drottningholm*, 54: 8–20.

Dobson, M.C. (1991). De-icing salt damage to trees and shrubs. *Forestry Commission Research Bulletin* 101, HMSO, London.

Dobson, S. and Cabridenc, R. (1990). *Tributyltin Compounds. Environmental Health Criteria No. 116*. World Health Organisation, Geneva.

DoE (1993). *Methodology for Identifying Sensitive Areas (Urban Waste Water Treatment Directive) and Methodology for Designating Vulnerable Zones (Nitrates Directive) in England and Wales. Consultation Paper*. Department of the Environment, London.

—— (1994) *Sustainable Development: The UK Strategy*. HMSO, London.

—— (1995). *A Guide to Risk Assessment and Risk Management for Environmental Protection*. HMSO, London.

—— (1996). *Indicators of Sustainable Development for the United Kingdom*. HMSO, London.

—— (1997). *UK National Air Quality Strategy*. Department of the Environment, London.

DoT (1994a). *Design Manual for Roads and Bridges, vol. 11, Environmental Assessment*. HMSO, London.

—— (1994b). *Safer Ships, Cleaner Seas. The Donaldson report*. Department of Transport, London.

Dubrova, Y.E., Nesterov, V.N., Krouchinsky, N.G., Ostapenko, V.A., Neuman, R., Neil, D. and Jeffreys, A.J. (1996). Human minisatellite mutation rate after the Chernobyl accident. *Nature*, 380: 683–6.

Dudley, N. (1990). *Nitrates – The Threat to Food and Water*. Green Imprint, London.

Dueck, T.A., Van der Eerden, L.J. and Berdowski, J.J.M. (1992). Estimation of SO_2 effect thresholds for heathland species. *Functional Ecology*, 6: 291–6.

Durant, D., Waddell, D.A., Benham, S.E. and Houston, T.J. (1992). *Air Quality and Tree Growth in Open Top Chambers. Research Information Note No. 208*. Forestry Commission, Farnham.

EA (1996a). *The Application of Toxicity-Based Criteria for the Regulatory Control of Wastewater Discharges. Consultation Paper*. Environment Agency, Bristol.

—— (1996b). *Understanding Buffer Strips*. Environment Agency, Bristol.

—— (1996c). *Validation of the UK–ADMS Dispersion Model and Assessment of its Performance Relative to R-91 and ISC Using Archived LIDAR Data*. Environment Agency, London.

—— (1997) *Best Practicable Environmental Option Assessment for Integrated Pollution Control Technical Guidance Note E1*. Environment Agency, Bristol.

EAC (1996). *Released Substances and their Dispersion in the Environment*. HMSO, London.

Eisler, R. (1995). Ecological and toxicological aspects of the partial meltdown of the Chernobyl power plant reactor. In D.J. Hoffman, B.A. Rattner, G.A. Burton and J. Cairns (eds) *Handbook of Ecotoxicology*. pp. 549–64. Lewis Publishers, Boca Raton.

Eisler, R. and Belisle, A.A. (1996). Planar PCB hazards to fish, wildlife, and invertebrates: a synoptic review. *Biological Report No. 31*. National Biological Service, Washington.

Ellenberg, H. (1987). Floristic changes due to eutrophication. In W.A.H. Asman and S.M.A. Dierden (eds) *Ammonia and acidification*. EURASAP, Bilthoven, the Netherlands.

ENDS (1993a). HMIP under fire over control of dioxins from Coalite works. *ENDS Report* 223: 6–7.

—— (1993b). Bolsover dioxin victims accept £200,000 from Coalite. *ENDS Report* 226: 4.

—— (1995a). Benefits of SO_2 abatement exceed costs, DoE study suggests. *ENDS Report* 241: 6.

—— (1995b). NRA lets Coalite off the hook over dioxin pollution. *ENDS Report* 246: 11–12.

—— (1996). Coalite fined record £150,000 for dioxin releases from incinerator. *ENDS Report* 253: 48–9.

English Nature (1994). *Water Quality for Wildlife in Rivers*. English Nature, Peterborough.

—— (1995). *Conservation in Catchment Management Planning – a Handbook*. English Nature, Peterborough.

EPAQS (1994a). *1,3-Butadiene. Report of the Expert Panel on Air Quality Standards*. HMSO, London.

—— (1994b). *Benzene. Report of the Expert Panel on Air Quality Standards*. HMSO, London.

—— (1994c). *Carbon monoxide. Report of the Expert Panel on Air Quality Standards*. HMSO, London.

—— (1994d). *Ozone. Report of the Expert Panel on Air Quality Standards*. HMSO, London.

—— (1995a). *Particles. Report of the Expert Panel on Air Quality Standards*. HMSO, London.

—— (1995b). *Sulphur Dioxide. Report of the Expert Panel on Air Quality Standards*. HMSO, London.

Erdman, J.A. and Modreski, P.J. (1984). Copper and cobalt in aquatic mosses and stream sediments from the Idaho Cobalt Belt. *Journal of Geochemical Exploration*, 20: 75–84.

Evans, P.A. and Ashmore, M.R. (1992). The effects of ambient air on a semi-natural grassland community. *Agriculture, Ecosystems and Environment*, 38: 91–7.

Everall, N.C. and Lees, D.R. (1996). The use of barley-straw to control general and blue-green algal growth in a Derbyshire reservoir. *Water Research*, 30: 269–76.

Everett, K.R. (1980). Distribution and properties of road dust along the northern portion of the haul road. In J. Brown and R. Berg (eds) *Environmental Engineering and Ecological Baseline Investigations Along the Yukon River–Purdhoe Bay Haul Road*, pp. 101–28. US Army Cold Regions Research and Engineering Laboratory, CRREL Report 80–19.

Farmer, A.M. (1990). The effects of lake acidification on aquatic macrophytes – a review. *Environmental Pollution*, 65: 219–40.

—— (1992). Catchment liming and nature conservation. *Land Use Policy*, 9: 8–10.

—— (1993a). SSSIs at risk from soil acidification in Britain. *JNCC Report* No. 156. Joint Nature Conservation Committee, Peterborough.

—— (1993b). The effects of dust on vegetation – a review. *Environmental Pollution*, 79: 63–75.

Farmer, A.M. and Bareham, S. (1993). The environmental implications of UK sulphur emission policy options for England and Wales. *JNCC Report No. 176*. Joint Nature Conservation Committee, Peterborough.

Farmer, A.M., Bates, J.W. and Bell, J.N.B. (1992). Ecophysiological effects of acid rain on bryophytes and lichens. In J.W. Bates and A.M. Farmer (eds) *Bryophytes and Lichens in a Changing Environment*, pp. 284–313. Oxford University Press, Oxford.

Forestry Commission (1993). *Forests and Water Guidelines* (third edn). Forestry Commission, Edinburgh.

Gee, A.S. and Stoner, J.H. (1989). A review of the causes and effects of acidification of surface waters in Wales and potential mitigation techniques. *Archives of Environmental Contamination and Toxicology*, 18: 121–30.

GESAMP (1990). *The state of the marine environment. Report by the UN Group of Experts on the Scientific Aspects of Marine Pollution*. Blackwell, Oxford.

—— (1991). Global strategies for marine environmental protection. *GESAMP Reports and Studies*, 45: 1–36.

Gibson, M.T., Welch. I.M., Barrett, P.R.F. and Ridge, I. (1990). Barley straw as an inhibitor of algal growth II: laboratory studies. *Journal of Applied Phycology*, 2: 241–8.

Gilbert, O.L. (1992). Lichen reinvasion with declining air pollution. In J.W. Bates and A.M. Farmer (eds) *Bryophytes and Lichens in a Changing Environment*, pp. 159–77. Oxford University Press, Oxford.

Glime, J. (1992). Effects of pollutants on aquatic species. In J.W. Bates and A.M. Farmer (eds) *Bryophytes and Lichens in a Changing Environment*, pp. 333–61. Clarendon Press, Oxford.

Goodwin, P.B. (1992). A review of new demand elasticities with special reference to short and long run effects of price changes. *Journal of Transport Economics and Policy*, 25: 155–69.

Gonzales, A.J. (1993). Global levels of radiation exposure: latest international findings. *IAEA Bulletin*, 4/1993: 49–51.

Greenpeace (1993). *The Poison Factory: The Story of Coalite Chemicals*. Greenpeace, London.

Harper, D. (1991). *Eutrophication of freshwaters*. Chapman and Hall, London.

Haslam, S.M. (1990). *River Pollution: an Ecological Perspective*. Belhaven Press, London.

Hawksworth, D.L. and Rose, F. (1970). Qualitative scale for estimating sulphur dioxide air pollution in England and Wales using epiphytic lichens. *Nature*, 227: 145–8.

Haycock, N. (1996). Constructed wetlands – can they cope? *Enact*, 4 (3): 17–20.

Hellawell, J.M. (1986). *Biological Indicators of Freshwater Pollution and Environmental Management*. Elsevier, London.

—— (1988). Toxic substances in rivers and streams. *Environmental Pollution*, 50: 61–85.

Henrikson, L., Hindar, A. and Thornelof, E. (1995). Freshwater liming. *Water, Air and Soil Pollution*, 85: 131–42.

Hillis, D.M. (1996). Life in the hot zone around Chernobyl. *Nature*, 380: 665–6.

HMIP (1996). *Decisions by the Chief Inspector and the Minister on Applications by the AGR and PWR Company Limited to Dispose of Radioactive Wastes from the Dungeness B, Hartlepool, Heysham 1, Heysham 2, Hinkley Point B and Sizewell Nuclear Power Stations*. HMIP, London.

Howells, G.D. (1984). Fishery decline: mechanisms and predictions. *Philosophical Transactions of the Royal Society*, 206 (B): 529–47.

ICE (1985). *A User Guide to Dust and Fume Control*. The Institute of Chemical Engineers, Rugby.

ILU (1995). *Annual Report 1995*. Institute of London Underwriters, London.

International Atomic Energy Agency (1992). *Effects of Ionizing Radiation on Plants and Animals at Levels Implied by Current Radiation Protection Standards*. Technical Report 332, IAEA, Vienna.

ITOPF (1987). *Response to Marine Oil Spills*. The International Tanker Owners Pollution Federation Ltd, London.

Jansen, E. and van Dobben, H.F. (1987). Is the decline of *Cantharellus cibarius* in the Netherlands due to air pollution? *Ambio*, 16: 211–13.

Kelly, M. (1988). *Mining and the freshwater environment*. Elsevier, London.

Kennedy, V.H., Horril, A.D. and Livens, F.R. (1990). *Radioactivity and Wildlife. Focus on Nature Conservation, 24*. Nature Conservancy Council, Peterborough.

Krester, W.A. and Colquhoun, J.R. (1984). Treatment of New York's Adirondack lakes by liming. *Fisheries*, 9: 36–41.

Landford, T.E.L. (1990). *Ecological Effects of Thermal Discharges*. Elsevier, London.

Laxen, D.P.H. and Thompson, M.A. (1987). Sulphur dioxide in Greater London, 1931–1985. *Environmental Pollution*, 43: 103–14.

Leck, C., Larsson, U., Bagander, L.E., Johansson, S., and Hajdu, S. (1990). Dimethyl sulfide in the Baltic Sea: annual variability in relation to biological activity. *Journal of Geophysical Research*, 95·C3: 3,353–63.

Lee, D.S. and Dollard, G.J. (1994). Uncertainties in current estimates of emissions of ammonia in the United Kingdom. *Environmental Pollution*, 86: 267–77.

Lee, J.A., Caporn, S.J.M. and Read, D.J. (1992). Effects on increasing nitrogen deposition and acidification on heathlands. In T. Schneider (ed) *Acidification, Research and Policy Applications*, pp. 97–106. Elsevier, Amsterdam.

Lee, J.A., Tallis, J.H. and Woodin, S.J. (1988). Acidic deposition and British upland vegetation. In M.B. Usher and D.B.A. Thompson (eds) *Ecological Change in the Uplands*, pp. 151–62. Blackwell, Oxford.

Leek, F. and de Savornin Lohman, A. (1996). Charges in water quality management in the EU. *European Environment* 6: 33–9.

Lovgren, K., Persson, G. and Thornelof, E. (1993). Acidification policy – Sweden. In T. Schneider (ed.) *Acidification Research, Evaluation and Policy Applications*. Elsevier, Amsterdam.

Lyons, M.G., Balls, P.W. and Turrell, W.R. (1993). A preliminary study of the relative importance of riverine nutrient inputs to the Scottish North Sea Coastal Zone. *Marine Pollution Bulletin*, 26: 620–8.

MacKenzie, S., Lee, J.A. and Wright, J.M. (1990). Ecological impact of liming blanket bog. *Report to the Nature Conservancy Council*, Peterborough.

McArdle, N. and Liss, P. (1996). The use of sulphur isotopes to distinguish between marine biogenic and anthropogenic sulphur in precipitation and aerosol in Wales. *Report to The Countryside Council for Wales*, Bangor.

McCallum, C. (1968). An epidemic of mussel poisoning in north-east England. *Lancet*, 2: 767.

Maddison, D., Pearce, D., Johansson, O., Calthorp, E., Litman, T. and Verhoef, E. (1996). *Blueprint 5: The True Costs of Road Transport*. Earthscan, London.

MAFF (1991). *Code of Good Agricultural Practice for the Protection of Water*. Ministry of Agriculture, Fisheries and Food, London.

—— (1992). *Code of Good Agricultural Practice for the Protection of Air*. MAFF, London.

—— (1996). *Radioactivity in food and the environment, 1995*. Ministry of Agriculture, Fisheries and Food, London.

Marsh, J.J. (1980). *Towards a Nitrate Balance for England and Wales*. Water Services, London.

Mason, C.F. (1991). *Biology of Freshwater Pollution*. Longman, London.

Mayerhofer, P., Weltschev, M., Trukenmuller, A. and Friedrich, R. (1995). A methodology for the economic assessment of material damage caused by SO_2 and NO_x emissions in Europe. *Water, Air and Soil Pollution*, 85: 2687–92.

Meybeck, M., Chapman, D. and Helmer, R. (1989). *Global Freshwater Quality*. Blackwell, Oxford.

Moss, B., Balls, H.R., Irvine, K. and Stansfield, J. (1986). Restoration of two lowland lakes by isolation from nutrient-rich water sources with and without removal of sediment. *Journal of Applied Ecology*, 23: 391–414.

Moss, B., Stansfield, J., Irvine, K., Perrows, M. and Phillips, G. (1996). Progressive restoration of a shallow lake: a 12 year experiment in isolation, sediment removal and bio-manipulation. *Journal of Applied Ecology*, 33: 71–86.

Murray, J.P., Harrington, J.J. and Wilson, R. (1982). Chemical and nuclear waste disposal. *The Cato Journal*, 2: 569–86.

NAPAP (1992). *1992 Report to Congress*. The National Acid Precipitation Assessment Programme, Washington.

National Academy of Sciences (1975). Marine litter. In Natural Research Council *Assessing Potential Ocean Pollutants*. Commission of Natural Resources. National Academy of Sciences, Washington.

National Research Council (1989). *Using Oil Dispersants on the Sea*. National Academy Press, Washington.

NCC (1991). *Nature Conservation and Pollution from Farm Wastes*. Nature Conservancy Council, Peterborough.

NETCEN (1996). *UK NO$_2$ Survey Annual Report*. AEA Technology, Harwell.

Newman, J. (1994). *Control of Algae With Straw. Information Sheet 3*. Aquatic Weeds Research Unit, Sonning.

Nicholson, J. (1996). The Coalite case. *HMIP Bulletin* 43: 7–8.

Nilsson, J. and Grennfelt, P. (1988). Critical loads for sulphur and nitrogen. *Nordisk Ministerrad Miljorapport, Oslo. 1988:* 16.

NPS (1988). *Air Quality in the National Parks. Natural Resources Report 88–1*. National Park Service, Washington.

NRA (1990). *Discharge consent and compliance policy: a blueprint for the future*. National Rivers Authority Water Quality Series No.1. NRA, Bristol.

—— (1992). *The Influence of Agriculture on the Quality of Natural Waters in England and Wales*. National Rivers Authority, Bristol.

—— (1993). *The Quality of the Humber Estuary 1980–1990. National Rivers Authority Water Quality Series No.12*. NRA, Bristol.

—— (1994). *Water Pollution Incidents in England and Wales – 1993*. National Rivers Authority, Bristol.

—— (1995). *Contaminants Entering the Sea. National Rivers Authority Water Quality Series No. 24*. NRA, Bristol.

NRPB (1996). *Radon Affected Areas: England and Wales*. National Radiological Protection Board, Didcot, Oxon.

Oksanen, J., Holopainen, J.K., Nerg, A. and Holopainen, T. (1996). Levels of damage of Scots pine and Norway spruce caused by needle miners along a SO$_2$ gradient. *Ecography*, 19: 229–36.

OPC (1993). *Nutrients in the Convention Area*. Oslo and Paris Commissions.

Ormerod, S.J. (1991). Acidification and freshwater biotas. In S.J. Woodin and A.M. Farmer (eds) *The Effects of Acid Deposition on Nature Conservation in Great Britain*, pp. 67–73. Nature Conservancy Council, Peterborough.

Ormerod, S.J. and Buckton, S.T. (1994). *Catchment Source Areas and the Ecological Effects of Wetland Liming – Phases I and II. R & D Note 281*, National Rivers Authority, Bristol.

Palo, R.T. and Wallin, K. (1996). Variability in diet composition and dynamics of radiocaesium in moose. *Journal of Applied Ecology*, 33: 1,077–84.

Perry, R. and McIntyre, A.E. (1986). Impact of motorway runoff upon surface water quality. In J.F. de L. G. Solbe (ed) *Effects of Land Use on Fresh Waters*, pp. 53–67. Ellis Horwood, Chichester.

Petterson, G. and Morrison, B. (1993). *Invertebrate Animals as Indicators of Acidity in Upland Streams. Forestry Commission Field Handbook 13*. Forestry Commission, Edinburgh.

Pinder, L.C.V., House, W.A. and Farr, I.S. (1993). Effects of insecticide on freshwater invertebrates. In A.S. Cooke (ed) *The Environmental Effects of Pesticide Drift*, pp. 64–75. English Nature, Peterborough.

Pitcairn, C.E.R. (1991). Effects of atmospheric nitrogen deposition on vegetation. In S.J. Woodin and A.M. Farmer (eds) *The Effects of Acid Deposition on Nature Conservation in Great Britain*. pp. 29–34. Focus on Nature Conservation No. 26. Nature Conservancy Council, Peterborough.

PORG (1993). *Ozone in the United Kingdom 1993*. Third Report of the United Kingdom Photochemical Oxidants Review Group. Department of the Environment, London.

QUARG (1996). *Airborne Particulate Matter in the UK. Report of the Quality of Urban Air Review Group*. Department of the Environment, London.

Radon Council (1995). *The Radon Manual*. The Radon Council, Shepperton.

Rainbow, P.S. (1987). Heavy metals in barnacles. In A.J. Southward (ed.) *Barnacle Ecology*, pp. 405–17. A. A. Balkema, Rotterdam.

Ramade, F. (1979). *Ecotoxicology*. J Wiley, Chichester.

RCEP (1992). *Sixteenth report of the Royal Commission on Environmental Pollution: Freshwater Quality*. HMSO, London.

—— (1994). *Transport and the Environment. Report of the Royal Commission on Environmental Pollution*, London.

Reiling, K. and Davison, A.W. (1992). The response of native, herbaceous species to ozone: growth and fluorescence screening. *New Phytologist*, 120: 29–37.

Renberg, I. and Hedberg, T. (1982). The pH history of lakes in SW Sweden as calculated from the subfossil diatom flora of the sediments. *Ambio*, 1: 30–3.

Richardson, D.H.S. (1975). *The Vanishing Lichens*. David & Charles, London.

Ridge, I. and Barrett, P.R.F. (1992). Algal control with barley straw. *Aspects of Applied Biology*, 29: 457–62.

Ritchie, W. and O'Sullivan (1994). *The Environmental Impact of the Wreck of the Braer. Report of the Ecological Steering Group on the Oil Spill in Shetland*. Scottish Office, Edinburgh.

Robinson, S. (1996). *Out of the Twilight Zone*. The Environment Council, London.

Roelofs, J.G.M., Kempers, A.J., Houdijk, A.L.F.M. and Jansen, J. (1985). The effect of airborne ammonium sulphate on *Pinus nigra* var. *maritima* in the Netherlands. *Plant and Soil*, 84: 45–56.

Rowell, A. (1996). *Green Backlash*. Routledge, London.

Ruhling, A. and Tyler, G. (1970). Sorption and retention of heavy metals in the woodland moss *Hylocomium splendens* (Hedw.) Br. et Sch. *Oikos*, 21: 92–7.

Sandhu, R. and Gupta, G. (1989). *Effects of NO$_2$ on growth and yield of Black Turtle Bean* (*Phaseolus vulgaris* L.) cv Domino. *Environmental Pollution*, 59: 337–44.

Satake, K., Nishikawa, M. and Shibata, K. (1989a). Distribution of aquatic bryophytes in relation to water chemistry of the acid river Akagawa, Japan. *Archiv für Hydrobiologie*, 116: 241–8.

Satake, K., Takamatsu, T., Soma, M., Shibata, K., Nishikawa, M., Say, P.J. and Whitton, B.A. (1989b). Lead accumulation and location in the shoots of the liverwort *Scapania undulata* (L.) Dum. in stream water at Greenside Mine, England. *Aquatic Botany*, 33: 111–22.

Satake, K., Soma, M., Seyama, H. and Uehiro, T. (1983). Assimilation of mercury in the liverwort *Jungermannia vulcanicola* Steph. in an acid stream Kashiranashigawa in Japan. *Archiv für Hydrobiologie*, 99: 80–92.

Savchenko, V.K. (1995). The ecological effects of the Chernobyl catastrophe. *Man and the Biosphere Series No. 16*. UNESCO, Paris.

SEEEC (1996). *Initial Report. Sea Empress Environmental Evaluation Committee*, Cardiff.

SEPA (1996). *A guide to Surface Water Best Management Practices*. Scottish Environmental Protection Agency, Stirling.

Skiba, U. (1991). Introduction to freshwater acidification. In S.J. Woodin and A.M. Farmer (eds) *The effects of Acid Deposition on Nature Conservation in Great Britain*, pp. 56–61. Nature Conservancy Council, Peterborough.

SOAFD (1992). *Scottish Office Agriculture and Fisheries Department Code of Good Practice for the Prevention of Environmental Pollution from Agriculture*. HMSO, London.

Sommer, S.G. and Hutchings, N. (1995). Techniques and strategies for the reduction of ammonia emission from agriculture. *Water, Air and Soil Pollution* 85: 237–48.

Steubing, L., Fangmeier, A., Both, R. and Frankenfield, M. (1989). Effects of SO$_2$, NO$_2$ and O$_3$ on population development and morphological and physiological parameters of native herb layer species in a beech forest. *Environmental Pollution*, 58: 281–302.

Sultanova, B.M. and Plisak, R.P. (1997). Degradation and peculiarities of vegetation restoration on the territory of Semipalatinsk nuclear test site (Republic of Kazakstan). In S. Kapur (ed.) *Proceedings of the First International Conference on Land Degradation*. (In Press).

Sutton, M.A., Place, C.J., Fowler, D. and Smith, R.I. (1994). *Assessment of the Magnitude of Ammonia Emissions in the United Kingdom*. Institute of Terrestrial Ecology, Penicuik, Midlothian.

TERG (1988). The effects of acid deposition on the terrestrial environment in the United Kingdom. *Report of the UK Terrestrial Effects Review Group*. HMSO, London.

Tyler, G. (1987). Probable effects of soil acidification and nitrogen deposition on the floristic composition of oak (*Quercus robur* L.) forest. *Flora*, 179: 165–70.

van der Eerden, L.J. and Duijm, N.J. (1988). An evaluation method for combined effects of SO$_2$ and NO$_2$ on vegetation. *Environmental Pollution*, 53: 468–70.

van der Eerden, L.J., Dueck, T.A., Berdowski, J.J.M., Greven, H. and van Dobben, H.F. (1991). Influence of NH$_3$ and (NH$_4$)$_2$SO$_4$ on heathland vegetation. *Acta Botanica Neerlandica*, 40: 281–96.

Van der Voet, E., Kleijn, R. and Udo de Haes, H. (1996). Nitrogen pollution in the European Union – origins and proposed solutions. *Environmental Conservation*, 23: 120–32.

Varley, M. (1967). British freshwater fishes: factors affecting their distribution. *Fishing News* (Books), London.

WCED (1987). *Our Common Future (The Brundtland Report). Report of the World Commission on Environment and Development*. Oxford University Press, Oxford.

Wellburn, A.R. (1990). Why are atmospheric oxides of nitrogen usually phytotoxic and not alternative fertilizers? *New Phytologist*, 115: 395–429.

Whicker, F.W. and Schultz, V. (1982). *Radioecology: Nuclear Energy and the Environment*, Volume 2. CRC Press, Boca Raton.

WHO (1995). *Update and revision of the Air Quality Guidelines for Europe*. Copenhagen, WHO.

Williams, C.T., Davis, B.N.K., Marrs, B.H. and Osborn, D. (1987). *Impact of Pesticide Drift*. NERC Report to the Nature Conservancy Council, Peterborough.

Woodin, S.J. and Farmer, A.M. (1993). Impacts of sulphur and nitrogen deposition on sites and species of nature conservation importance in Great Britain. *Biological Conservation*, 63: 23–30.

Woodwell, G.M. and Gannutz, T.P. (1967). Effects of chronic gamma irradiation on lichen communities of a forest. *American Journal of Botany*, 54: 1,210–15.

Yocum, J.E. and Baer, N.S. (1984). Chapter E-7, section 7.1 In: *The Acidic Deposition Phenomenon and its Effects: Critical Assessment Review Papers. EPA Report 600/8-83-016BF*. US Environmental Protection Agency, Washington.

Zannetti, P. (1992). *Air Pollution Modelling*. Van Nostrand Reinhold, New York.

Index

Aberdeen 208
abstraction 104, 111
Accelerated Vehicle Retirement Scheme 94
acid rain 22, 32, 40, 66, 79, 144, 150, 180
acid mists 40
acidification 108, 136
acute effects 34
Adirondack Mountains 138, 144
Adriatic Sea 176
agent orange 32
agriculture 12, 27, 61, 73, 89, 111, 112, 113, 118, 124, 126, 136, 147, 150, 157, 159, 176, 179, 181, 198
Air Quality Management Areas 68
Alaska 169, 218
Alkali Acts 58
allergens 35, 38
aluminium 40, 137, 150
americium 193, 195, 203
ammonia 22, 24, 44, 61, 66, 73, 76, 85, 89, 99, 100, 128
ammonium 106, 127
amphibians 195
Anglian Water 122
angling 215
animal housing 90
animal waste 90
anti-foulants 159, 163
Arctic 32, 203
arsenic 159, 160, 162, 164
Artemisia frigida 197
Artemisia scoparia 197

Asellopsis intermedia 182
asthma 15, 34, 35, 37, 38
Athens 28
Atlantic Ocean 177, 181
Austria 92, 108

1,3 Butadiene 64, 82
bacteria 107, 112, 119, 146, 195
badgers 209
Baltic Sea 166, 176, 180, 201
Bangkok 39
barnacles 159, 163, 171
bathing *see* swimming
Bathing Water Directive 107, 183
BATNEEC 13, 60, 68, 73
bats 172
Belarus 198
Belgium 177
benzene 15, 64, 72, 82, 92
best practicable means (BPM) 58
bioaccumulation 11, 159, 161, 194
biodiversity 4
biofuels 92
biological oxygen demand 118, 128
biomonitoring 51, 138, 149, 163
birds 149, 170, 171, 195
black list 106
Black Triangle 31
blood flukes 114
blooms (phytoplankton) 125, 130, 135
blue-green algae 118, 126, 128
Boston 29

BPEO 60, 70, 73, 223
Braer 170
Branchiura sowerbyi 146
Brazil 92, 98
bream 132
Brittany 169
Broughton Island 32
Bryophytes 66, 116; *see also* mosses and
     liverworts
buffer strips 136, 145, 182
buildings 8, 51, 54, 211
Brent Spar 223

caddis-flies 138
cadmium 38, 106, 116, 157, 158, 159, 160,
     161, 162, 164
caesium-137 194, 195, 200, 202, 203
Cairo 39
calcium 137, 194
California 36, 93
Calor Group 223
Canada 30, 32, 34, 92
cancer 122, 149, 193, 195, 209
Carbaryl 147
carbon monoxide 24, 35, 64, 82, 88, 93, 94
carcinogens 3
cardiovascular disease 34
carp 129, 132, 146
cars *see* transport
catalytic converters 25, 26, 74, 92, 93, 94
catchment management plans 110
cattle 90, 113, 119
Centroptilum pennulatum 148
cetaceans 170
chemical dispersants 171
chemical release inventory 61
Chernobyl 11, 17, 191, 193, 196, 198, 205
Cheviot Hills 210
China 30, 31, 223
Chironomus 119
chlorine 114
chlorofluorocarbons 223
cholera 114
chromium 106, 159, 161, 162, 164
chromosomes 200
chronic effects 34
Chrysochondromulina 178
civil service 222
Clean Air Act (US) 35, 54
Clean Air Acts (UK) 59
Coalite Chemicals 61
Coastwatch UK 158
cold war 190
Common Agricultural Policy 182

company cars 98
common scoter 172
contingent valuation 54
contraceptive pill 149
Control of Pollution Act (1974) 105
copper 116, 118, 144, 157, 159, 160, 161, 162,
     164
coral reefs 167, 176
Cornwall 176, 209, 210
cosmic radiation 210
crabs 204
critical group 207, 208
critical levels 28, 66, 77, 99
critical loads 40, 76, 77, 100, 142
crops 47, 207
crustaceans 195, 204
Cryptosporidium 114
Cumbria 203, 208, 209, 210
cyanide 106
cyclones 29, 88
Czech Republic 31

Dangerous Substances Directive 106, 160
Danube (river) 201
Daphnia 134, 147
DDT 106, 165
Deltamethrin 147
Denmark 177, 180
depuration 175
Derbyshire 61
Derris 147
Devon 114, 210
diatoms 124, 139, 178
diesel 16, 30, 92
diffuse pollution 108, 158, 182
dimethyl sulphide 180
dinoflagellates 178
dioxins 3, 61, 81, 88, 149, 166, 216
dippers 139
direct toxicity assessment 109, 185
dispersion (including modelling) 7, 9, 69, 83
DNA 194
dogwhelks 163
Donaldson Report 172
Dounreay 193
Dover Strait 173
Drepanocladus fluitans 118
drinking water 106, 110, 112, 122, 123, 149,
     175, 207
Drinking Water Directive 107
dumping from ships 175
dunes 172
Dungeness 208
dust 22, 29, 48, 85, 195

East Anglia 29
Eastern Europe 75
economics 11, 13, 34, 54, 69, 70, 75, 123, 134,
    167, 169
Eichornia crassipes 126, 127
Ekofisk 167
electrostatic precipitator 29, 89
Elodea canadensis 119
employment 206
endocrine disruptors 148
English Nature 132
Entamoeba 112, 114
Enteromorpha 119, 176
Environment Act (1995) 59, 68, 71, 94, 109
Environment Agency (UK) 17, 60, 68, 69,
    109, 110, 112, 136, 150, 185, 207, 208
Environment Service (Northern Ireland) 60
environmental assessment 99
environmental assessment levels 69
environmental oestrogens 148
Environmental Protection Act (1990) 11, 13,
    59, 67
Environmental Protection Agency (US) 53,
    209
epiphytes 44, 126
Eristalis tenax 119
Escherichia coli 114, 175
ethanol 92
ethinyl oestrodiol 149
European Environment Agency 17, 71
European Union 70, 105; *see also separate*
    directives
eutrophication 123, 176
Exxon Valdez 169, 218

Farne Islands 178
fertilizer 90, 91, 127, 150, 176
Fifth Environmental Action Programme 108
filter bags 88
Finland 108
Firth of Clyde 182
fish 120, 126, 129, 131, 137, 146, 149, 158,
    161, 162, 169, 195, 196, 200, 207, 204
fish farming 111, 113, 118, 124, 178
fisheries 110, 111, 170
fishing nets 157
flags of convenience 172
Florida 126, 176, 177
flue-gas cleaning 87, 89
fluoride 8, 25, 45, 85
Fontinalis antipyretica 118
forestry 111, 113, 144, 198
forests 40, 66
Framework Directive (water) 106

France 108, 169, 177
freshwater 77, 104–53
Friends of the Earth 223
frost 44
fugitive emissions 58, 81
fur seals 157

Gammarus pulex 148
garbage 157
genetic damage 200
genetics 11
Gennada valens 204
German Bight 175
Germany 31, 68, 73, 95, 177, 180
Giarda lamblia 114
Gloucestershire 211
Great Ouse (river) 122
green parties 224
Greenfreeze 223
greenhouse gases 90, 92
Greenpeace 62, 223, 224
grey list 106
groundwater 24, 122, 149, 181
guillemots 170, 172
Gulf of Mexico 167
Gymnodinium 178

half-life 191
Hartlepool 208
health 14, 15, 34, 54, 64, 71, 74, 99, 108, 112,
    122, 149, 161, 175, 176, 183, 194, 202,
    204, 208, 219
heart disease 15
heathlands 40
heavy metals 26, 38, 76, 115, 120, 128, 158,
    183, 200
Helsinki Convention 184
hepatitis 112, 114
Her Majesty's Nuclear Installations
    Inspectorate 207
Her Majesty's Inspectorate of Pollution 60
Heteropappus altaicus 197
Heysham 208
Hinkley Point 204, 208
Humber (estuary) 160, 163
Hydrilla verticilata 126
hydrochlorofluorocarbons 223

Imposex 17, 163
India 47, 51
industry 222
integrated combined cycle gassification 87
Integrated Pollution Control 12, 58, 59, 73,
    106, 163, 182, 183

Integrated Pollution Prevention and Control
    Directive 3, 72, 108, 183
intelligence quotient 15, 38
International Commission on Radiological
    Protection 206
invertebrates 119, 120, 128, 130, 138, 143,
    146, 147, 161, 195
iodine-131 199, 200
ion-exchange resins 127
IQ *see* intelligence quotient
Irish Sea 166, 203
iron 116, 118
Italy 93, 203
Ixtoc 167

Japan 94, 161
Jungermannia vulcanicola 118

Kazakstan 197
klondikers 172
Kochia scoparia 197
Kuwait 167

landfill 116, 118, 124, 128, 131, 133, 150, 158,
    181
Laplanders 203
LD$_{50}$ 10
lead 15, 38, 64, 71, 74, 107, 116, 118, 121,
    157, 159, 160, 161, 162, 164, 204
Lecanora conizaeoides 51
Lee (river) 122
leeches 119
Lemna minor 119, 126
lichens 49, 67, 85, 143, 195, 196, 202
limestone 89
liming 139
Lincolnshire 211
liquid petroleum gas 92
litter 182
liverworts 143
lobsters 204
Local Agenda 21 (LA21) 218
local authority air pollution control 60, 67
Local Environment Agency Plans 110
logging 224
London 28, 34, 53, 58, 68, 93, 98, 114
Los Angeles 28, 93, 96
low-NO$_x$ burners 26, 88

Maastricht Treaty 108
mammals 195
manganese 116
mangroves 167
Marine Environmental High Risk Areas 172

Marpol Convention 184
mayflies 119, 120, 138, 219
media 219
medical radiation 192, 210
meningitis 114
mercury 106, 116, 118, 158, 159, 161, 162
metals 89; *see also* heavy metals
methane 90
methanol 92
Mexico 176
Mexico City 29, 98
mining 111, 116, 158, 224
Ministry of Agriculture, Fisheries and Food
    208
mitochondria 200
modelling 83
Mogilev 200
molluscs 119, 149, 161, 167, 175, 195, 204
molybdenum 116, 144
monitoring 6, 9, 16, 71, 76, 79, 83, 100, 115,
    149, 150, 173, 182, 185, 196, 203, 204,
    219
moose 203
mosses 85, 195
MOT test 94
Municipal Waste Water Treatment Directive
    108
mycorrhizae 44
Myriophyllum spicatum 126, 146

National Air Quality Strategy (UK) 23, 63,
    68, 69, 98
National Radiological Protection Board 206,
    209
National Rivers Authority 60, 109, 132
National Society for Clean Air 223
natterjack toads 144
natural gas 30, 92
nature conservation 78
Netherlands 177
New York State Electric Company 120
nickel 38, 106, 159, 161, 162, 164
Nitrate Sensitive Areas 122
Nitrate Vulnerable Zones 122
nitrates 24, 107, 108, 122, 124, 176
Nitrates Directive 108, 123, 183
nitrogen deposition 44
nitrogen dioxide/monoxide (NO$_x$) 15, 17, 22,
    25, 26, 34, 35, 47, 51, 54, 55, 64, 65, 66,
    71, 72, 74, 76, 82, 83, 85, 88, 89, 93, 94,
    95, 96, 99, 100, 180
nitrous oxide 90
no effect level (NOEL) 9
non-governmental organisations (NGOs) 75

Norfolk Broads 129, 134
North Sea 175, 178, 179, 180, 181
North Sea Conference 181, 184
Northamptonshire 211
Norway 75, 138, 141, 142, 177, 178, 180, 203
Novay Zemlya 197
Nuclear Installations Act (1965) 208
nuclear power 192, 203, 207
nuclear tests 192, 196, 197
nuclear waste 203
nuclear weapons 191

occult deposition 33
oil 121, 157, 166, 182, 184, 218
organic pollution 118, 120, 175
Organisation for Economic Cooperation and
    Development 124
organochlorines 165
Oslo 98
Oslo Convention 177, 184
otters 138, 169, 170, 219
ozone 14, 28, 34, 36, 47, 51, 53, 64, 66, 71,
    81, 82, 83, 100, 114

Panama 167
paralytic shellfish poisoning 178
Paris 96
Paris Convention 177, 184
particulates 22, 25, 29, 34, 36, 48, 52, 53, 55,
    65, 74, 81, 88, 93, 94, 95
pathogens 112, 183
peatlands 40, 143
percentiles 23
persistent organic toxins 76
pesticides 147, 149, 161, 165
petrol 16, 30, 38
Phaeocystis 176
Philadelphia 209
phosphates 108, 123, 124, 128, 132, 176
photochemical oxidant creation potential 32
Phragmites 128, 198
phytoplankton 129, 176, 195, 204
pigs 90, 113, 119
pilchards 173
pine trees 195, 196, 200
plutonium 38, 193
PM$_{10}$ 15, 29, 36, 64, 82
Poecilia reticulata 146
Poland 31, 74
polio 114
politicians 221
polonium 204
polyaromatic hydrocarbons 30, 31
polychlorinated biphenyls 30, 31, 149, 165

Potamogeton crispus 119
Potamogeton pectinatus 119, 126
Potamogeton perfoliatus 119, 146
poultry 90, 113
precautionary principle 6
prescribed substances 73
press releases 220
Protection of Groundwater Directive 106
Psatyrostachys juncea 197
public 75, 215, 221
Public Health Act (1937) 105
publicity 169
puffins 172

quarrying 29, 49

rabbits 203
radioactive waste 184, 193
Radioactive Substances Act (1993) 207
radioactivity 190–212
radon 3, 208
red grouse 143
red list 106
red tides 177
red-throated diver 172
reed beds 128
Regional Seas Programme 184
reindeer 202
reptiles 149, 195
respiratory disease 34
Rhine States Conference 181
Rio Earth Summit 4
risk management 205
River Pollution Commission 105
River Pollution Prevention Act (1876) 205
roads see transport
Round Loch of Glenhead 139
roundworms 114
Royal Society for the Protection of Birds 224
Russia 198, 203
ruthenium 203
Rutland Water 128

Sagittaria 119
salmon 146, 170
Salmonella 114
salt marshes 171, 172
sand eels 170
Scandinavia 40, 73, 202
Scapania undulata 118
Schistosoma 112, 114
Scotland 138, 139, 179, 193, 210
Scottish Environmental Protection Agency
    60, 68

Sea Empress 169, 171, 219
seafood 175
seals 166, 170
secondary pollutants 14, 28
secondary treatment 175
sediment 145
seeder-feeder effect 34
selective catalytic reduction 89
selective non-catalytic reduction 89
Sellafield 203, 208, 209
Semipalatinsk 197
Seveso 32
sewage 61, 113, 114, 116, 118, 119, 127, 131,
    133, 149, 157, 158, 163, 174, 176, 179,
    183, 219
sewage fungus 118
shags 170
sheep 203
shellfish 160, 162, 169, 171, 175, 207
Shellfish Waters Directive 183
Shetland Islands 170, 173
Shigella 114
shipwrecks 158
shrimps 204
Sigara dorsalis 148
silage 124, 129
Sites of Special Scientific Interest 131
Sizewell 208
slurry 119
smog 27, 28, 31, 34, 53, 68
smoke 35, 71
smokeless zones 59
smoking 38
smoky vehicles hotline 94
snails 137
snow 34
Snowdonia 203
social values 3
soil 40, 77
solvents 32, 150
Somerset 211
song thrushes 203
Soviet Union 197, 198
Spain 203, 217
Sparganium 119
Sphaerotilus natans 118
Sphagnum 138, 143
standards 2, 8, 9, 13, 35, 36, 37, 38, 63, 65,
    71, 73, 74, 109, 123, 162, 183, 204, 206,
    209
Statutory Water Quality Objectives 112
steppe 197
Stipa sareptana 197
stoneflies 119, 120, 138

Stour (river) 122
straw 130
Strontium-90 194, 201, 203
Sullom Voe 173
sulphur oxides 15, 22, 25, 30, 34, 38, 47, 49,
    51, 53, 54, 55, 56, 59, 64, 65, 66, 71, 72,
    76, 81, 82, 83, 87, 89, 99, 100
Surface Waters Directive 107
Surfers Against Sewage 175, 176
suspended solids 114, 119, 121, 127, 128,
    130, 145
sustainable development 4, 205, 218
Sweden 51, 92, 108, 136, 138, 139, 159, 177,
    178, 180, 201, 203
swimming 126, 175, 176, 207
Swindon 114
Switzerland 177

Taenia 114
tainting 169, 170
Taiwan 116
Taj Mahal 51
tapeworms 114
Tay (river) 179
terrorism 206
tertiary butyl ether 74
Thailand 92
Thames (river) 105, 114, 122, 159
thermal pollution 3, 145, 182
thoron 210
thyroid 200
tin 160
Torrey Canyon 169, 173
tourism 169
transport 14, 25, 26, 30, 32, 48, 52, 68, 71, 72,
    74, 81, 91, 113, 120, 157, 161, 180, 216
tree health 42, 44, 47, 67
tributyl tin 17, 159, 162
Trichuris 114
tritium 195
trout 120, 138, 146, 147
Tweed (river) 179
Typha angustifolia 198
typhoid 114, 199

Ukraine 198
Ulva 176
UN Convention on the Law of the Sea 183
uncertainty 7
United Nations Economic Commission for
    Europe 75
United States of America 35, 40, 53, 64, 92,
    94, 224
units of measurement 23, 191

unleaded petrol 16, 74, 92
uranium 193, 204, 208, 209
Urban Waste Water Treatment Directive
    132, 183

Valisneria spiralis 146
Vibrio cholera 114
Vietnam 32
violence 224
viruses 195
visibility 53, 54
volatile organic compounds 15, 22, 25, 32, 71,
    76, 81
volcanic activity 30, 116
voles 200, 201

Wales 138, 144, 169, 171, 180, 210, 219

war 206
waste 81, 184, 205
wastewater 12
Water Act (1989) 109
Water Resources Act (1991) 112
white-tailed eagle 159
willingness-to-pay 14
Wise Use Movement 224
World Wide Fund for Nature (WWF) 224
World Health Organisation 64
World Bank 74

York 58
Ythan (river) 179

zinc 116, 121, 157, 158, 159, 160, 162, 164
zooplankton 126, 129, 134